COLLINS REFERENCE

DICTIONARY OF STATISTICS

COLLINS
REFERENCE

DICTIONARY OF STATISTICS

ROGER PORKESS

COLLINS
London and Glasgow

First published 1988

Diagrams by Gordon Barr

ISBN 0 00 434354 9

British Library Cataloguing in Publication Data

Porkess, R.
Collins dictionary of statistics.
1. Mathematical statistics—Dictionaries
I. Title
519.5'03'21 QA276.14

Printed in Great Britain by Collins, Glasgow

To Ketaki

CONTENTS

PREFACE

In recent years there has been a marked increase in the teaching and application of statistics. A generation ago the subject was rarely taught in schools and very little in universities. Nowadays every school child learns at least some descriptive statistics within the mathematics syllabus; nearly all A-level mathematics courses now contain some probability and statistics topics which for many students, comprise the applied part of their syllabus. In addition, many other subjects, such as geography and biology require some knowledge of statistical techniques from their A-level students. At the tertiary level, the importance of statistical evidence is now accepted by virtually all academic disciplines and there are relatively few science or social science subjects which do not require knowledge of the subject. The increasing use of data bases allows easy access to data and the use of computers takes much of the tedium out of complex calculations.

At the same time public awareness of the subject has increased. For example, it is now expected that when the result of an opinion poll is announced the sample size should be given. Statistical terms, such as parameter and correlation have found their way into everyday English, often used incorrectly.

This rapid growth of the subject means that many people are now using it with a somewhat limited theoretical or mathematical background and it is hoped that such people, be they at work, school or university will find this book particularly helpful.

The content of the Dictionary has been selected to cover the needs not only of those taking statistics as part of A-level, or similar, mathematics courses but also of those learning it as a service subject at the tertiary level. While its format is superficially like that of a traditional dictionary, it differs in a number of respects. Many of the entries are encyclopedic in treatment and include many worked examples. These illustrate how the various statistical measures are calculated and the tests applied. Most of these examples use artificial data; real data is usually more difficult to work with and can easily obscure the point being made. The text is illustrated throughout with diagrams and graphs to aid the reader.

Considerable effort has been made to ensure that the information in this book is accessable to its readers. Care has been taken to avoid using terminology that is likely to cause difficulty to the reader of any particular entry. The simpler the subject of an entry, the easier the language and notation used within it. On the other hand, more advanced topics can often only be understood in terms of easier ones and so there is something of a hierarchy among the entries. There is extensive cross-referencing throughout the book, indicated by the use of SMALL CAPITALS.

The conventions used are those in common usage at the time of writing. Greek letters refer to parent population parameters, Roman letters to sample statistics. Capital letters are used for variable names, lower case for values of those variables. A full list of symbols is included in Appendix A while Appendix B is a list of formulae. The tables in Appendix C are those needed for the tests and techniques described within the book.

Acknowledgements

I would like to thank the many people who have helped this book along its way – colleagues, editors, advisors, etc.; in particular Dr Kevin McConway of the Open University and Mr Alan Downie of Imperial College who advised on the text and checked it for accuracy; Janet France for her editorial work on it, including setting up all the cross-referencing; and to Ian Crofton, Edwin Moore and James Carney from Collins Reference Division for their work in turning the typescript into this book.

Roger Porkess

A

abscissa, *n.* the horizontal or *x*-coordinate in a two-dimensional system of Cartesian coordinates. The vertical or *y*-coordinate is called the *ordinate*. See Fig. 1.

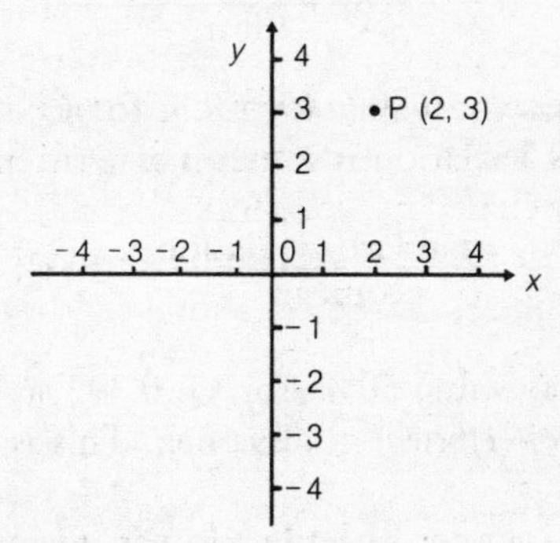

Fig. 1. **Abscissa.** In this example, the abscissa of P is 2, and the ordinate of P is 3.

absolute frequency, see FREQUENCY.

absolute measure of dispersion, *n.* a measure of DISPERSION (or spread) relative to the size and distribution of the figures involved.

The sets of numbers P and Q

P	2	4	6	8	10	12	14
Q	1002	1004	1006	1008	1010	1012	1014

both have the same RANGE (12), INTERQUARTILE RANGE (8) and STANDARD DEVIATION (4). Relative to the sizes of the numbers, however, P is clearly much more spread than Q. If, for example, the figures represented typical incomes of people in two countries (in suitable units of currency), they would be remarkably uniform in Q but not at all so in P.

There are several absolute measures:

(a) *Coefficient of mean deviation.* This is given by:

$$\frac{\text{mean absolute deviation}}{\text{mean}}$$

For the examples given above, this has values 3.43/8 = 0.43 for P, and 3.43/1008 = 0.0034 for Q. A similar coefficient can be found using the MEDIAN or MODE instead of the mean.

(b) *Coefficient of variation.* This is given by:

$$\frac{\text{standard deviation}}{\text{mean}}$$

For P, this has the value 4/8 = 0.5; for Q, 4/1008 = 0.00397. Coefficient of variation is sometimes written as a percentage:

$$\frac{\text{standard deviation}}{\text{mean}} \times 100\%$$

For P, this has value 50%; for Q, 0.397%.

(c) *Quartile coefficient of dispersion.* This is given by

$$\frac{\text{upper quartile} - \text{lower quartile}}{\text{upper quartile} + \text{lower quartile}}$$

For P, the QUARTILES are 12 and 4; for Q, 1012 and 1004. Thus the values of the quartile coefficient of dispersion are:

$$\text{for P,} \quad \frac{12-4}{12+4} = 0.5$$

$$\text{for Q,} \quad \frac{1012-1004}{1012+1004} = 0.00397$$

absolute value or **modulus,** *n.* the magnitude of a real number or quantity, regardless of its sign. Symbol: straight brackets | |.

$$|-6| = 6 \qquad |17.3| = 17.3$$

acceptance number or **allowable defects,** *n.* the greatest number of defectives in a sample that will allow the batch (from which the sample is drawn) to be accepted without further quality-control testing. Symbol: n. In a SINGLE SAMPLING SCHEME for QUALITY CONTROL, a sample is taken from each batch of the product. If the number of defectives is no larger than the acceptance number, the whole batch is accepted.

addition law, *n.* a law concerning the probabilities of two events occurring. The addition law states that:

$$p(A \text{ or } B) = p(A) + p(B) - p(A \text{ and } B)$$

It is often written, using SET notation, as:

$$p(A \cup B) = p(A) + p(B) - p(A \cap B)$$

Example: if a card is drawn at random from a pack of 52 playing cards, what is the probability that it is a club or a queen? (See Fig. 2).

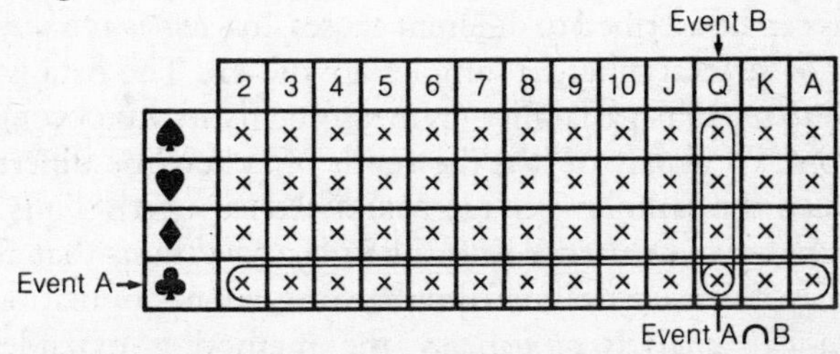

Fig. 2. **Addition law.** Picking a card at random from the pack, Event A∩B is where either a club or a queen is drawn. The probability of Event A∩B is 4/13.

Event A: The card is a club. Probability = 1/4.
Event B: The card is a queen. Probability = 1/13.
Event $(A \cap B)$: The card is the queen of clubs.
Probability = 1/52.
Event $(A \cup B)$: The card is a club or a queen.

$$\begin{aligned} p(A \cup B) &= p(A) + p(B) - p(A \cap B) \\ &= 1/4 \quad + 1/13 - 1/52 = 16/52 = 4/13 \end{aligned}$$

The probability that the card is a club or a queen is 4/13.

age, *n.* **1.** the period of time that a person, animal or plant has lived, or that an object has existed, i.e. the *chronological age*. **2.** the level in years that a person has reached in any area of development compared with the normal level for his chronological age, for example, *mental age*, *reading age*.

aggregate, *n.* the value of a single variable resulting from the combination of data for a number of variables. For example, the cost-of-living index is an aggregate of the various components that are used to form it. An examination mark is often an aggregate of the marks of several different papers, and perhaps a project. The process of forming an aggregate is called *aggregation*.

alienation, *n.* the effect of sampling errors on a CORRELATION COEFFICIENT.

allowable defects, see ACCEPTANCE NUMBER.

alternative hypothesis, see NULL HYPOTHESIS.

analysis of variance or **anova,** *n.* a technique for investigating how much of the variability in a set of observations can be ascribed to different causes. In a *one-way analysis of variance*, several different samples are drawn. The data is then tested to see if its variability (as measured by its VARIANCE) is all random, or if part of it is the result of systematic differences between the samples (see FACTORIAL EXPERIMENT). Thus one-way analysis of variance tests the NULL HYPOTHESIS that all the samples have been drawn from the same parent population. (In a *two-way analysis of variance* the method is extended to variation for two factors.)

Analysis of variance involves separating the sample variance into two components, that within the samples and that between them. These are then compared using the F-TEST. The variance is separated using the SUM OF SQUARES result:

$$\sum x_{\mathrm{T}}^2 = \sum x_{\mathrm{B}}^2 + \sum x_{\mathrm{W}}^2$$

Total sum of squares = That between the groups + That within them.

Example: a class of 24 small children was divided into three groups, each of which was then taught reading by a different method. After 3 months the children were tested, using a 0–10 scale. Their marks were:

Group P : 5, 0, 3, 5, 5, 4, 8, 2
Group Q: 4, 5, 4, 7, 5, 10, 3, 10
Group R : 3, 5, 0, 3, 3, 9, 4, 9

Do these marks support the theory that there is no difference in the results from the three methods, at the 5% significance level? Null hypothesis: All the figures have been drawn from the same parent population.
The calculations are set out below, using the convention that the symbol x is used for an actual score and X for its deviation.

Table of marks for the three samples and their squares.

	x_P	x_P^2	x_Q	x_Q^2	x_R	x_R^2
	5	25	4	16	3	9
	0	0	5	25	5	25
	3	9	4	16	0	0
	5	25	7	49	3	9
	5	25	5	25	3	9
	4	16	10	100	9	81
	8	64	3	9	4	16
	2	4	10	100	9	81
Total $\sum$	32	168	48	340	36	230

The sums of squares are calculated using

$$\sum X^2 = \sum x^2 - (\sum x)^2/n$$

P : $$\sum X_P^2 = 168 - \frac{32^2}{8} = 40$$

Q: $$\sum X_Q^2 = 340 - \frac{48^2}{8} = 52$$

R : $$\sum X_R^2 = 230 - \frac{36^2}{8} = 68$$

$$\sum x = 32 + 48 + 36 = 116 \quad \sum x^2 = 168 + 340 + 230 = 738$$

Total (T): $\sum X_T^2 = 738 - \frac{116^2}{24} = 117.3$

Between groups (B): $\sum X_B^2 = \frac{32^2}{8} + \frac{48^2}{8} + \frac{36^2}{8} - \frac{116^2}{24} = 17.3$

Within groups (W): $\sum X_W^2 = 40 + 52 + 68 = 160$

Check: $\sum X_T^2 = \sum X_B^2 + \sum X_W^2$

$177.3 = 17.3 + 160$

The between-groups variance ($\hat{\sigma}_B^2$) for the number of groups k, is estimated as

$$\hat{\sigma}_B^2 = \frac{\sum X_B^2}{k-1} = \frac{17.3}{3-1} = 8.65$$

The within-groups variance ($\hat{\sigma}_W^2$) for n items, is estimated as

$$\hat{\sigma}_W^2 = \frac{\sum X_W^2}{n-k} = \frac{160}{24-3} = 7.62$$

From this, the ratio F is found:

$$F = \frac{\sigma_B^2}{\sigma_W^2} = \frac{8.65}{7.62} = 1.14$$

At the 5% significance level, $F_2^{21} = 19.44$.
Since $1.24 < 19.44$, the null hypothesis is accepted. On the basis of this test, there is no evidence to suggest that it makes any difference which method is used to teach children to read.

Analysis of variance may be carried out on the ranks of the data, rather than the actual values; see KRUSKAL–WALLIS ONE-WAY ANALYSIS OF VARIANCE, FRIEDMANN'S TWO-WAY ANALYSIS OF VARIANCE BY RANK.

anova, *abbrev. for* ANALYSIS OF VARIANCE.

antilogarithm, see LOGARITHM.

AOQ, see AVERAGE OUTGOING QUALITY.

AOQL, see AVERAGE OUTGOING QUALITY LIMIT.

approximation, *n.* the process or result of making a rough calculation, estimate or guess. Approximation is used extensively when deciding if an answer is reasonable. The statement 'It takes five hours to fly from Britain to Australia by jumbo jet' is clearly false. The distance is about 20,000km, the speed about 1,000km per hour, and so the time taken is about 20 hours. To decide that the statement was false, it was not necessary to know exact values of the speed of the aeroplane or the distance involved.

When giving numerical information, the number of significant figures should be consistent with the accuracy of the information involved. Thus the statement 'The cost of running my car is £1,783.47 per year' is misleading because it is not possible to estimate it so accurately, especially when depreciation is taken into account. A better figure would have been 'approximately £1,800', which is obviously a round number, indicating an error of perhaps £50 either way.

arithmetic mean, *n.* the result obtained by adding the numbers or quantities in a set and dividing the total by the number of members in the set. It is often called AVERAGE. For example, the arithmetic mean of 43, 49, 63, 51 and 28 is:

$$\frac{43+49+63+51+28}{5}=46.8$$

The arithmetic mean of the numbers $x_1, x_2, \ldots x_n$ is given by:

$$\frac{x_1+x_2+\ldots+x_n}{n}$$

This may also be written as:

$$\frac{1}{n}\sum_{i=1}^{n} x_i \quad \text{or} \quad \frac{\sum_{i=1}^{k} x_i f_i}{\sum_{i=1}^{k} f_i}$$

where f_i is the frequency of the value x_i, and there are k distinct values of x. See also MEAN.

association, *n.* the tendency of two events to occur together. If the probabilities (or frequencies) of the different combinations of events A and not A, and B and not B, are

	A	*not A*
B	*a*	*b*
not B	*c*	*d*

the association is said to be positive if $ad > bc$, and negative if $ad < bc$. If $ad = bc$, the events A and B are independent. *Yule's coefficient of association* is given by

$$\frac{ad - bc}{ad + bc}$$

At times the term association is also applied to variables, but it is more usual to use the term CORRELATION when the value of one variable tells you something about the value of another.

assumed mean, *n.* an estimated or approximate value for the ARITHMETIC MEAN, or average, used to simplify its calculation.

When working out the arithmetic mean of a set of numbers of similar size, the calculation can sometimes be simplified by:

(a) Assuming a mean;

(b) Working out the differences from it (+ or −) for each number;

(c) Calculating the mean of the differences;

(d) Adding the mean of the differences onto the assumed mean.

It does not matter what value is taken for the assumed mean, but the nearer it is to the true mean, the smaller are the numbers involved.

Example: find the mean of the ages of six children who are, in years and months, 13y11m, 14y4m, 13y6m, 14y6m, 14y1m and 14y2m,

(a) Take an assumed mean: 14y0m;

(b) Calculate the differences from the assumed mean: −1, +4, −6, +6, +1 and +2 months;

(c) Calculate the mean of the difference:

$$\frac{-1 + 4 - 6 + 6 + 1 + 2}{6} = +1 \text{ month}$$

(d) the mean age of the children is thus given by:

true mean = assumed mean + mean of differences
= 14y0m + 1m

The mean age of the children is 14y1m.

attribute testing, *n.* testing a product for quality control when the result is either 'good' or 'defective'. Attribute testing may be contrasted with VARIABLE TESTING where the result is a quantitative measure, like the length of a nail or the resistance of an electrical component. See also LATTICE DIAGRAM.

average, *n.* **1.** a measure of central tendency, the ARITHMETIC MEAN. **2.** (in certain circumstances) other types of mean. Examples are given under GEOMETRIC MEAN and HARMONIC MEAN. **3.** (in everyday use) typical or normal. Statements are made like, 'The average cost of a laying hen is three pounds', or 'William is average at mathematics'. In this context, it has come to be used as any measure of central tendency; thus MODE and MEDIAN can also be regarded as types of average. See also MOVING AVERAGE.

average outgoing quality (AOQ), *n.* the proportion of defective items being sent out with a product after a quality-control scheme has been implemented. For a single inspection scheme (SINGLE SAMPLING SCHEME) with ACCEPTANCE NUMBER n from samples of size N, with 100% inspection of failed batches, AOQ is given by

$$\text{AOQ} = \text{P(p)}\ (1 - \text{n/N})\text{p}$$

where p is the proportion of defectives manufactured, and P(p) is the probability of a batch with proportion p defective being accepted.

average outgoing quality limit (AOQL), *n.* the highest average proportion of defective items that will be let through in the long run, following a quality-control sampling scheme.

When AVERAGE OUTGOING QUALITY (AOQ) is plotted against the proportion of defectives manufactured, p, the graph rises to

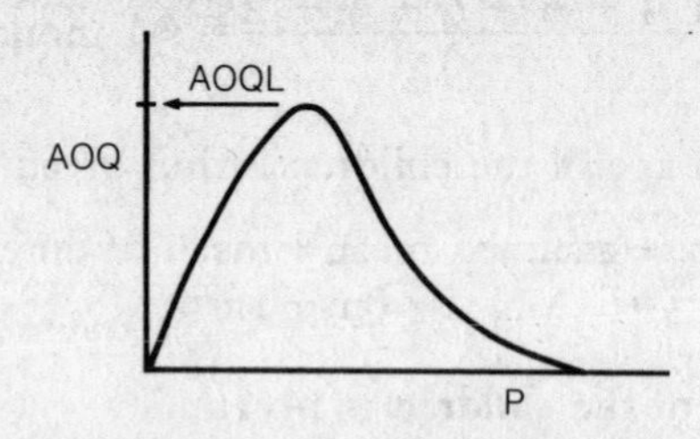

Fig. 3. **Average outgoing quality limit (AOQL).** The AVERAGE OUTGOING QUALITY (AOQ) of goods is plotted against the proportion of defective goods (p) to find the maximum value (AOQL).

a maximum and then falls away. A typical graph for a SINGLE SAMPLING SCHEME is shown in Fig. 3. If p is small, the AOQ will be low because there are few defectives to get through. If p is large, the sampling scheme is likely to result in most batches being stopped and undergoing 100% inspection; any defectives are then rejected. Between these two cases are values of p which are neither very small, nor large enough to involve most batches being stopped.

The maximum value shown on such a graph is called the average outgoing quality limit, and represents the highest average proportion of defectives that will be let through in the long run. It is, however, possible for worse individual batches to get through.

B

band chart, *n.* a percentage BAR CHART.

bar chart or **bar diagram,** *n.* a display of data using a number of rectangles, of the same width, whose lengths are proportional to the frequencies they represent. Bar charts can be drawn in several ways; for example, with the bars drawn vertically or horizontally (Figs. 4, 5), immediately next to each other, or with a fixed gap between them. A *component* or *compound bar chart* has two or more parts to each bar (see Fig. 6). There are many ways in which such diagrams can be drawn, some of them very artistic. See also POPULATION PROFILE, THREE-QUARTERS HIGH RULE. Compare VERTICAL LINE CHART.

bar diagram, see BAR CHART.

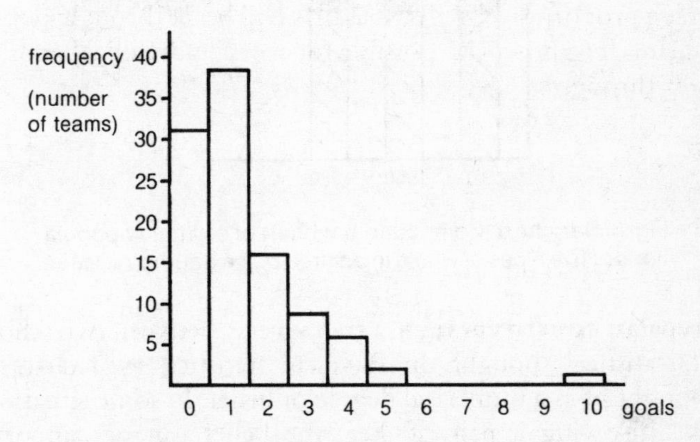

Fig. 4. **Bar chart.** Goals scored in the World Cup finals in 1982, in Spain.

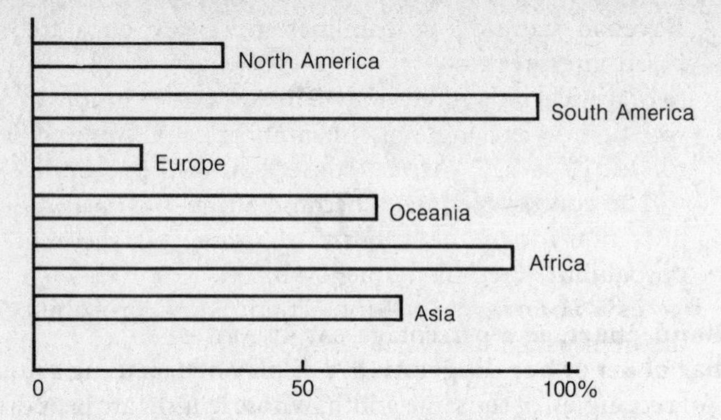

Fig. 5. **Bar chart.** Percentage increases in population, 1950–1980. (Source: *Philip's Certificate Atlas*, 1978).

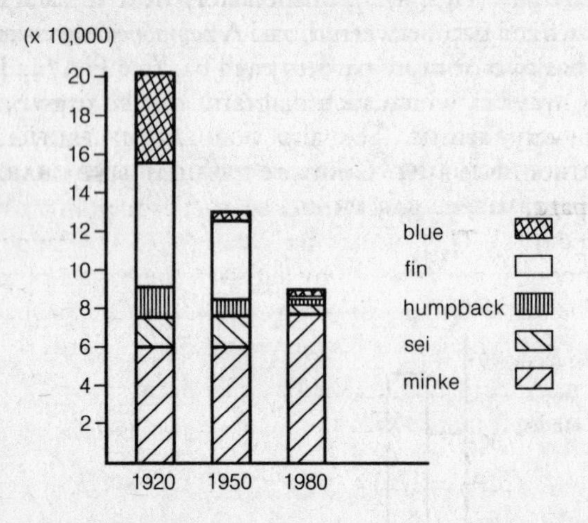

Fig. 6. **Bar chart.** Compound bar chart of estimated populations of five types of whale in an area of the Southern Ocean.

Bayesian controversy, *n.* a controversy between two schools of statistical thought. In Bayesian statistics, PROBABILITY is thought of as a numerical degree of belief. In some situations, like throwing a pair of dice, the belief can be supported objectively. In others, such as predicting the price of gold next week, any probability used is essentially subjective. In non-

Bayesian statistics, probabilities must rely on a frequency-based interpretation.

Statistical inference involves drawing conclusions about the situation underlying experimental data, and this may involve the use of BAYES' THEOREM on conditional probability.

The controversy arises because a non-Bayesian statistician does not accept the validity of using this theorem if the probabilities are not frequency-based.

Bayes' theorem, *n.* (on conditional probability), the theorem which states that

$$p(A \mid B) = \frac{p(B \mid A)\, p(A)}{p(B \mid A)\, p(A) + p(B \mid A')\, p(A')}$$

where $p(A \mid B)$ is the probability of event A occurring, given that event B has occurred, and where A' is the event 'not A'. Thus Bayes' theorem relates $p(A \mid B)$ to $p(B \mid A)$ and $p(B \mid A')$. Example: all the men from an African village go on a leopard hunt on a special day every year. It is a somewhat perilous occasion with a probability of 1/10 that a man is attacked by a leopard; when one is attacked the probability of his being killed is 1/3. There are several other sources of danger, so that if someone is not attacked by a leopard he has a 1/60 probability of being killed on the hunt. What is the probability that someone who is killed is the victim of a leopard?

Event A: A man is attacked by a leopard.
Event B: A man is killed.

$$p(A) = 1/10 \qquad p(B \mid A) = 1/3$$

$$p(A') = 9/10 \qquad p(B \mid A') = 1/60$$

Then Bayes' theorem gives:

$$p(A \mid B) = \frac{\frac{1}{3} \times \frac{1}{10}}{\frac{1}{3} \times \frac{1}{10} + \frac{1}{60} \times \frac{9}{10}} = \frac{20}{29}$$

So the probability that a man who died was the victim of a leopard is 20/29.

Bayes' theorem can be written in more general form as

$$p(A_r \mid B) = \frac{p(B \mid A_r) \times p(A_r)}{\sum_{r=1}^{r=n} p(B \mid A_r) \times p(A_r)}$$

where $A_1 \ldots A_n$ are mutually EXCLUSIVE and EXHAUSTIVE events.

Bernoulli distribution, *n.* another name for BINOMIAL DISTRIBUTION. The term is, however, usually restricted to the special case of the binomial distribution when n = 1.

Number of successes:	0	1
Probability:	1 − p	p

Thus it is the PROBABILITY DISTRIBUTION of the number of successes in a single trial.

Bernoulli's theorem, *n.* the theorem which states that, if the probability of success in a trial is p, the probability of exactly r successes in n independent trials is:

$$^{n}C_{r} p^{r} (1-p)^{n-r}$$

where $^{n}C_{r} = \frac{n!}{(n-r)!\,r!}$ and n! (n factorial) is $1 \times 2 \times 3 \times \ldots \times n$. (See also BINOMIAL COEFFICIENTS.)

This is the theorem underlying the BINOMIAL DISTRIBUTION.

Bernoulli trial, *n.* an experiment with fixed probability p of success, 1 − p of failure. In a sequence of Bernoulli trials, the probability of success remains constant from one TRIAL to the next, independent of the outcome of previous trials.

The probability distribution for the various possible numbers of successes, 0 to n, in n Bernoulli trials is the BINOMIAL DISTRIBUTION. This can also be called the BERNOULLI DISTRIBUTION, but this term is often restricted to the case when n = 1.

beta distribution, *n.* the distribution with probability density function given by:

$$f(x) = \frac{x^{a-1}(1-x)^{b-1}}{B(a, b)}$$

where the parameters $a, b > 1, 0 \leqslant x \leqslant 1$, and $B(a, b)$ is the *beta function* given by

$$B(a, b) = \int_0^1 x^{a-1}(1-x)^{b-1}\,dx$$

The beta distribution has

$$Mean = \frac{a}{(a+b)}$$

$$Variance = \frac{ab}{(a+b+1)(a+b)}$$

The graph of the beta distribution varies in shape according to the values of a and b, as shown in Fig. 7.

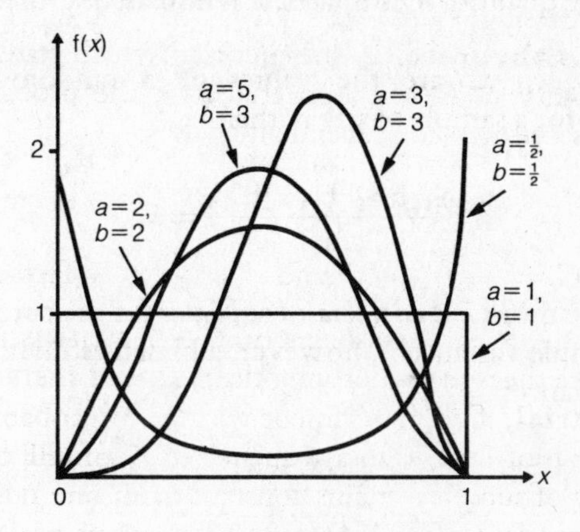

Fig. 7. **Beta distribution.**

beta function, see BETA DISTRIBUTION.

bias, *n.* a systematic ERROR in a statistical result. Errors from chance will cancel each other out in the long run, but those from bias will not.

Example: the police decide to estimate the average speed of drivers using the fast lane of a motorway, and consider how it can be done. One method suggested is to tail cars, using a police patrol car, and record their speeds as being the same as that of the police car. This is likely to produce a biased estimate, as any driver exceeding the speed limit will slow down on seeing a police car coming up behind. The police then decide to use an unmarked car for their investigation; this should eliminate that source of bias. If, however, the meter of the car they use does not indicate correct values, their estimate of drivers' speeds will still be subject to bias.

biased estimator, *n.* a formula or procedure which produces a biased estimate of a PARAMETER of the parent population from sample data. In statistics it is common to estimate properties of a parent population, like its MEAN and VARIANCE, by examining a sample. A formula or procedure which predicts a property or parameter of the parent population using the data is called an ESTIMATOR. If its expected value over all possible samples is equal to the quantity it estimates, it is unbiased. Otherwise it is biased.

If $x_1, x_2, \ldots, x_n$ are the values of a random variable measured, for a sample of size n, then

$$\frac{x_1 + x_2 + \ldots + x_n}{n} = \bar{x}$$

The sample mean $\bar{x}$ is an *unbiased estimator* for the parent mean.

The sample variance is, however, a biased estimator for the parent variance:

$$\sum_{i=1}^{n} \frac{(x_i - \bar{x})^2}{n}$$

An unbiased estimator for it is:

$$\sum_{i=1}^{n} \frac{(x_i - \bar{x})^2}{n-1}$$

bimodal, *adj.* (of a distribution) **1.** having two MODES. Example: the goals scored per match by a football team in a season are given in the table below. There are two modes, 0 and 2, so the distribution is bimodal.

Goals/match	*Frequency (matches)*
0	12
1	8
2	12
3	6
4	3
5	1

2. a distribution which has two distinct, but unequal, local modes, as in Fig. 8.

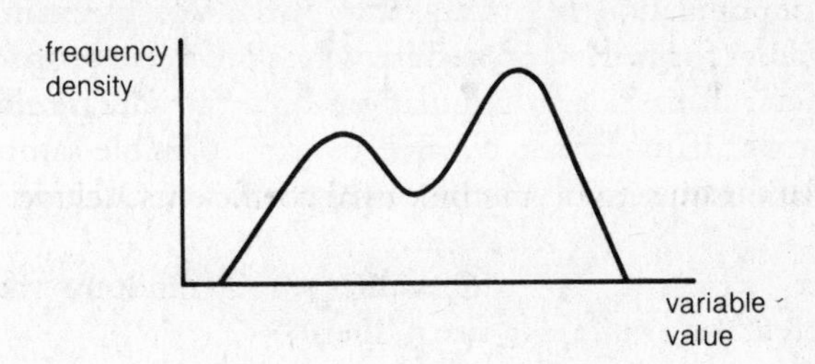

Fig. 8. **Bimodal.** A bimodal distribution.

binomial coefficients, *n.* the coefficients of the terms in the expansion of $(x+y)^n$ for positive integer values of n. Symbol: ${}^nC_r$.

Example:

$$(x+y)^5 = 1x^5 + 5x^4y + 10x^3y^2 + 10x^2y^3 + 5xy^4 + 1y^5$$

and so the binomial coefficients for n = 5 are 1 5 10 10 5 1.

There are two common notations for writing these:

(a) ${}^5C_0$, ${}^5C_1$, ${}^5C_2$, ${}^5C_3$, ${}^5C_4$ and ${}^5C_5$

(b) $\binom{5}{0}, \binom{5}{1}, \binom{5}{2}, \binom{5}{3}, \binom{5}{4}$ and $\binom{5}{5}$.

There are two ways of working out binomial coefficients.

(a) Use of the formula ${}^nC_r = \dfrac{n!}{r!(n-r)!}$

Thus $^5C_2 = \frac{5!}{2!3!} = \frac{5.4.3.2.1}{2.1.3.2.1} = 10$

(b) PASCAL'S TRIANGLE:

$$\begin{array}{ccccccccccccc} & & & & & & 1 & & & & & & \\ & & & & & 1 & & 1 & & & & & \\ & & & & 1 & & 2 & & 1 & & & & \\ & & & 1 & & 3 & & 3 & & 1 & & & \\ & & 1 & & 4 & & 6 & & 4 & & 1 & & \\ & 1 & & 5 & & 10 & & 10 & & 5 & & 1 & \\ 1 & & 6 & & 15 & & 20 & & 15 & & 6 & & 1 \end{array}$$

● ● ● ● ● ● ● ●

Relationships involving binomial coefficients include

$$^nC_r = {}^nC_{n-r}$$
$$^nC_r + {}^nC_{r+1} = {}^{n+1}C_{r+1}$$
$$^nC_0 + {}^nC_1 + {}^nC_2 + \ldots + {}^nC_n = 2^n$$

^nC_r also gives the number of ways of selecting r objects from n, without regard to their order.

binomial distribution, *n.* the distribution of the probabilities of the various possible numbers of successes (0 to n) when n independent BERNOULLI TRIALS are carried out. Symbol: B(n, p). See also BERNOULLI DISTRIBUTION. Compare PASCAL'S DISTRIBUTION. If the probability of success is p, and that of failure $q = 1 - p$, then the probability of r successes and $n - r$ failures is given by:

$$^nC_r p^r q^{n-r} \quad \left(\text{the binomial coefficient } ^nC_r = \frac{n!}{r!\,(n-r)!}\right)$$

The PROBABILITY GENERATING FUNCTION for the binomial distribution is given by $(q + pt)^n$ since, when expanded, this gives:

$$^nC_0 q^n + {}^nC_1 q^{n-1}(pt) + \ldots + {}^nC_r q^{n-r}(pt)^r + \ldots + {}^nC_n (pt)^n$$

It is often convenient to write the expression to be expanded as $(pX + qY)^n$, where event X has probability p, event Y, i.e. X′, or not X, has probability $q = 1 - p$. X and Y can then be thought of as markers, so that the probability of X occurring r times, and Y occurring $n - r$ times, is given by the coefficient of $X^r Y^{n-r}$, namely:

$$^nC_{n-r}p^r q^{n-r} = \frac{n!}{(n-r)!\,r!}p^r q^{n-r}$$

The MEAN, or expectation, of the binomial distribution is given by np, the STANDARD DEVIATION by $\sqrt{npq}$. (See also FACTORIAL.)

Example: in a certain country, 60% of people support the Democratic Party, the rest the Parliamentary Party.

(a) Five people are selected at random. What are the probabilities that 0, 1, 2, 3, 4 and 5 of them support the Parliamentary Party?

(b) Opinion pollsters interview people in sets of five. What are the mean and standard deviation of the numbers of Parliamentary Party supporters in these sets?

(a) The distribution is generated by the expression $(.6D + 4P)^5$, giving:

Democratic Party	*Parliamentary Party*		*Probability*
5	0	$^5C_0(.6)^5$	.0778
4	1	$^5C_1(.6)^4(.4)$	.2592
3	2	$^5C_2(.6)^3(.4)^2$	.3456
2	3	$^5C_3(.6)^2(.4)^3$	.2304
1	4	$^5C_4(.6)(.4)^4$	.0768
0	5	$^5C_5(.4)^5$	.0102

(b) *Mean* $= np = 5 \times .4 = 2$

Standard deviation $= \sqrt{npq} = \sqrt{5 \times .4 \times .6} = 1.095$

The binomial distribution can be approximated by (a) the NORMAL DISTRIBUTION, mean np and standard deviation $\sqrt{npq}$, when n is large and p is neither small nor near 1, and (b) by the POISSON DISTRIBUTION, mean and variance both np, when n is large and p small.

The distribution of the proportion of successes in n trials is binomial with

$$Mean = p$$

$$Standard\ deviation = \sqrt{\frac{pq}{n}}$$

biometry, *n.* the study of biological data by means of statistical analysis.

birth rate, *n.* the ratio of live births per unit time in a specified area, group, etc., to population, usually expressed per thousand population per year. In the 1870s, the birth rate in the UK was about 35 per 1,000, compared to 11.8 in 1977, the lowest recorded figure.

biserial, *adj.* (of the relationship between two variables) having one variable that takes only two values.

bivariate distribution, *n.* a distribution of two random variables Compare MULTIVARIATE DISTRIBUTION.

Example: a second-hand car, of a particular make and model, has two easily measurable variables of interest to a prospective buyer: its age and the mileage it has covered. (There are also a number of other factors, like its general condition and the driving habits of the previous owner.) A dealer, buying a batch of 30 cars of a particular model from a company which runs a fleet of them, tabulates the information as a bivariate frequency or CONTINGENCY TABLE, as in Fig. 9.

In this example the two variables, age and mileage, are not independent; the older cars have usually covered a greater distance. It would, consequently, not make much sense to consider the distributions of the two variables separately; they are better looked at together as a bivariate distribution. The level of association between the two variables can be judged visually if a SCATTER DIAGRAM is drawn, or calculated as a CORRELATION COEFFICIENT.

If the probabilities of the two variables are given, rather than their frequencies, the distribution is then a *bivariate probability*

		MILEAGE (× 1000 miles)							
		0–10	*10–20*	*20–30*	*30–40*	*40–50*	*50–60*	*60–70*	*70–80*
	0–1	1				1	1		
	1–2					2			
	2–3		1	1	2	2	1		
AGE	*3–4*					1	2	5	1
(years)	*4–5*						2		1
	5–6		1	1				1	
	6–7								1
	7–8				1	1			

Fig. 9. **Bivariate distribution.** The bivariate frequency table of the information compiled by a second-hand car dealer about a batch of 30 cars.

distribution. The distributions obtained by adding along each row or column of a bivariate distribution table are called *marginal distributions*.

In the example of the second-hand cars (Fig. 9), the marginal distributions are:

Mileage (× *1,000*)	0–10	10–20	20–30	30–40	40–50	50–60	60–70	70–80
Frequency	1	2	2	3	7	6	6	3

and

Age	0–1	1–2	2–3	3–4	4–5	5–6	6–7	7–8
Frequency	3	2	7	9	3	3	1	2

If each row (or column) of a bivariate distribution table has the same variance, the distribution is said to be *homoscedastic* with respect to that variable, otherwise *heteroscedastic*.

bivariate frequency table, see CONTINGENCY TABLE.

block, *n.* a subset of the items under investigation in a statistical experiment. This could be:
The barley in a particular part of a field.
The householders in one street of a town.
The weekly earnings of one particular employee of a factory.
The variability within a block would be expected to be less than that within the whole population being investigated.

box-and-whisker plot, *n.* a diagram used to display some important features of the distribution of experimental data (see EXPLORATORY DATA ANALYSIS). It consists of two lines (whiskers) drawn from points representing the extreme values to the lower and upper quartiles, and a box drawn between the quartiles which includes a line for the median. A box-and-whisker plot is also called a *boxplot*.
Example: the population densities (people/ha) of the English non-metropolitan counties:

Avon	6.8	Humberside	2.4
Bedfordshire	4.0	Isle of Wight	3.0
Berkshire	6.3	Kent	3.9
Buckinghamshire	2.7	Lancashire	4.5
Cambridgeshire	1.7	Leicestershire	3.3
Cheshire	3.9	Lincolnshire	0.9
Cleveland	6.7	Norfolk	0.7
Cornwall and Isles of Scilly	1.2	Northamptonshire	2.2
		Northumberland	0.6
Cumbria	0.7	North Yorkshire	0.8
Derbyshire	3.4	Nottinghamshire	4.4
Devon	1.4	Oxfordshire	2.1
Dorset	2.2	Shropshire	1.0
Durham	2.5	Somerset	1.2
East Sussex	3.6	Staffordshire	3.3
Essex	3.9	Suffolk	1.5
Gloucestershire	1.9	Surrey	5.9
Hampshire	3.8	Warwickshire	2.4
Hereford and Worcester	1.5	West Sussex	3.1
Hertfordshire	5.8	Wiltshire	1.5

For these figures, (source: *The Geographical Digest*, 1978.)
Greatest value = 6.8 (Avon)
Upper quartile = 3.9 (Cheshire, Essex or Kent)
Median = 2.7 (Buckinghamshire)
Lower quartile = 1.5 (Hereford and Worcester, Suffolk or Wiltshire)
Least value = 0.6 (Northumberland)

The boxplot is drawn as in Fig. 10. The box, drawn between the lower and upper quartiles, contains 50% of the data. The two whiskers, from the quartiles to the least or greatest values, each contain 25% of the data.

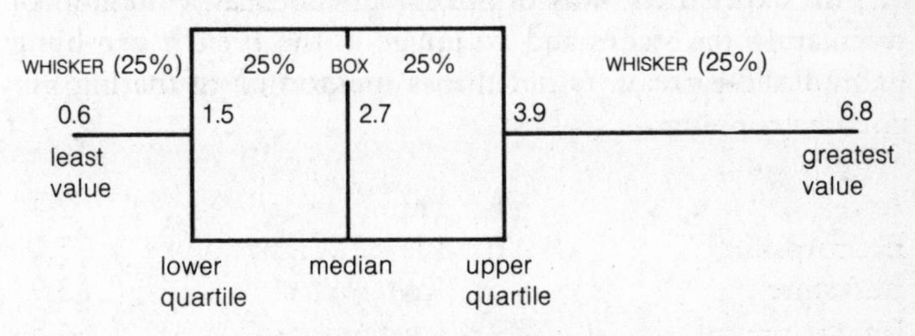

Fig. 10. **Box-and-whisker plot.**

boxplot, see BOX-AND-WHISKER PLOT.

Buffon's needle, *n.* a statistically-based experiment for determining π. A piece of paper is marked with parallel lines, distance a apart, and a needle is thrown onto the paper in a random manner (see Fig. 11). The probability that the needle crosses one of the lines is given by:

$$\frac{2l}{a\pi}$$

where l is the length of the needle, and $l < a$. Thus if the needle is dropped a large number of times, it is possible to estimate π as

$$\frac{2l}{a} \times \frac{\text{Total number of throws}}{\text{Number of landings crossing a line}}$$

This experiment is a very slow way of finding π. After 10,000 throws, the first decimal place should be known with

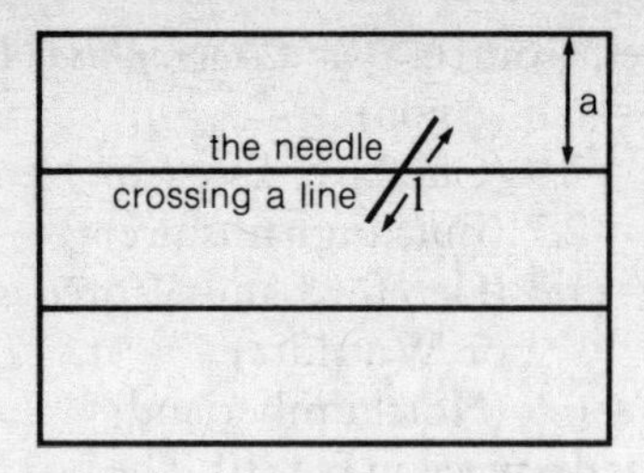

Fig. 11. **Buffon's needle.**

reasonable confidence, the second not at all. It is also very difficult to ensure a random throw of the needle.

This experiment was originally intended as a method of calculating the stakes and winnings in the French gambling game franc-carreau, rather than a method of estimating π.

C

cardinal number, *n.* a number denoting quantity but not order in a group. Thus 1, 2, 3, . . . are cardinal numbers. The word cardinal is used in contrast to *ordinal* which expresses position in a sequence, e.g. first, second, third etc. (see ORDINAL SCALE).

Cartesian graph, see GRAPH, ABSCISSA.

categorical scale or **nominal scale,** *n.* a scale which sorts variables according to category. Thus, the cars in a park can be sorted according to make, the fish in a pond according to type, etc. For a categorical scale it is essential that the categories be EXHAUSTIVE, EXCLUSIVE and clearly defined. Thus every variable should belong to one and only one category, and there should be no doubt as to which one.

There are two types of categories, a priori and constructed. *A priori categories* already exist, for example, a person's sex or year of birth; *constructed categories* are decided upon by research workers. A bad choice of categories can prejudice the outcome of an investigation. Categories may be indexed by numbers which do not correspond to any true order, like the numbers on footballer's shirts.

causation, *n.* the production of an effect by a cause. Causation can never be proved statistically, although it may be very strongly suggested.

A common misunderstanding concerns the relationship between CORRELATION and causation. If there is a strong correlation between the variables measuring the occurrence of two events A and B, it may be that

A causes B, $A \Rightarrow B$ or B causes A, $B \Rightarrow A$
or they both cause each other, $A \Leftrightarrow B$

but it may also be that a third event C causes both A and B,

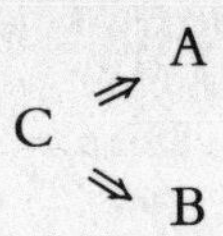

or that a more complex set of interactions is going on. It may also be that the correlation observed is merely due to chance fluctuations in the variables involved and that there is no causal reason for it at all. The statement 'Correlation does not imply causation' describes this situation.

The fact that there is a strong (negative) correlation between the mean number of cars per family and the mean number of children per family for the various countries of the world does not mean that owning a car prevents you having children, or that having children stops you having a car. The situation is more complicated when the affluence of the country is involved; in affluent countries, birth rates are low and car ownership is common.

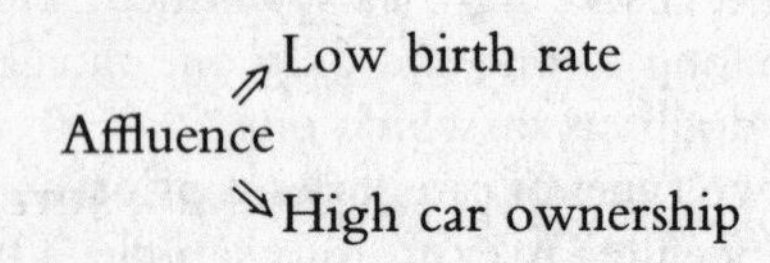

(It could, however, be argued that affluence and low birth rates both cause each other.)

cause variable, see EXPLANATORY VARIABLE.

census, *n.* an official periodic count of a population, including such information as sex, age, occupation, etc. In the UK a census is held every ten years (1961, 1971, 1981, . . .). Heads of household are required to provide information about all those living under their roofs. The questions asked cover name, age, family status, education, and employment.

A mini-census is sometimes taken in the interval between censuses using, typically, a 10% sample of the population.

central limit theorem, *n.* the theorem which states that, if samples of size n are taken from a parent population with mean μ and standard deviation σ, then the distribution of their means will be approximately normal, with

$$Mean = \mu$$

$$Standard\ deviation = \frac{\sigma}{\sqrt{n}}$$

As the sample size n increases, this distribution approaches the normal DISTRIBUTION with increasing accuracy (see SAMPLING DISTRIBUTION). Thus in the limit, as $n \to \infty$, the distribution of the sample means → Normal, mean μ, standard deviation $\sigma/\sqrt{n}$.

If the parent population is itself normal, the distribution of the sample means will be normal, whatever the sample size. If the parent population is of finite size N, two possibilities arise:

(a) If the sampling is carried out with replacement, the theorem stands as stated;

(b) If there is no replacement, the standard deviation of the sample means is:

$$\frac{\sigma}{\sqrt{n}}\sqrt{\frac{N-n}{N-1}}$$

The factor $\sqrt{\frac{N-n}{N-1}}$ is called the FINITE POPULATION CORRECTION.

The central limit theorem provides the basis for much of sampling theory; it can be summed up, as follows. If n is not small, the sampling distribution of the means is approximately normal, has mean $= \mu$ (the parent mean), and has standard deviation $= \sigma/\sqrt{n}$ (where σ is the parent standard deviation).

certainty, *n.* the condition of an event whose probability of occurrence is equal to 1.

chance, *n.* **1.** the unknown and unpredictable element that causes an event to have one outcome rather than another. **2.** that which is described by probability.

characteristic function, see MOMENT-GENERATING FUNCTION.

Charlier's check, *n.* an accuracy check that can be used when calculating MEAN and STANDARD DEVIATION. It uses the fact that, for a set of numbers $x_1, x_2, \dots x_n$, with frequencies $f_1, f_2, \dots f_n$,

$$\sum_{i=1}^{n} f_i (x_i + 1)^2 = \sum_{i=1}^{n} f_i x_i^2 + 2 \sum_{i=1}^{n} f_i x_i + \sum_{i=1}^{n} f_i$$

and is illustrated in the following example.
Example: Find the standard deviation of

5, 5, 4, 4, 4, 3, 3, 2, 1, 1

x_i	f_i	$f_i x_i$	$f_i x_i^2$	Extra column $f_i(x_i+1)^2$
5	2	10	50	72
4	3	12	48	75
3	2	6	18	32
2	1	2	4	9
1	2	2	2	8
	$\sum f_i = 10$	$\sum f_i x_i = 32$	$\sum f_i x_i^2 = 122$	$\sum f_i(x_i+1)^2 = 196$

The extra column $f_i(x_i+1)^2$ is worked out and used to check $\sum f_i\ (=n)$, $\sum f_i x_i$ and $\sum f_i x_i^2$, all of which would normally be calculated anyway. In this case

$$\sum f_i(x_i+1)^2 = \sum f_i x_i^2 + 2 \sum f_i x_i + \sum f_i$$
$$196 = 122 + 2 \times 32 + 10$$

which is easily seen to be true. It is then a simple matter to complete the calculations of mean and standard deviation, secure in the knowledge that the figures are correct.

$$Mean = \frac{\sum f_i x_i}{\sum f_i} = \frac{32}{10} = 3.2$$

$$Standard\ deviation = \sqrt{\left\{\frac{\sum f_i x_i^2}{\sum f_i} - \text{mean}^2\right\}}$$
$$= \sqrt{\left\{\frac{122}{10} - 3.2^2\right\}} = 1.4$$

chart, see GRAPH.

Chebyshev's inequality, *n.* the theorem which states that: If a probability distribution has mean μ and standard deviation σ, then the probability that the value of a random variable with that distribution differs from μ by more than $k\sigma$ is less than $1/k^2$, i.e.

$$\Pr\{|x-\mu|>k\sigma\}<\frac{1}{k^2}$$

for any $k>0$.

chi-squared (or χ^2) distribution, *n.* the distribution, with n degrees of freedom, of $\chi^2 = X_1^2 + X_2^2 + \ldots + X_n^2$, where X_1, X_2, ... X_n are n independent normal random variables with mean 0 and variance 1. Because the χ^2 distribution is different for each value of n (see Fig. 12, where n=7) complete tables would be very cumbersome; they are therefore usually given only for CRITICAL VALUES for certain SIGNIFICANCE LEVELS (e.g., .99, .95, .90, .10, .05, .02, .01; see tables on p. 249).

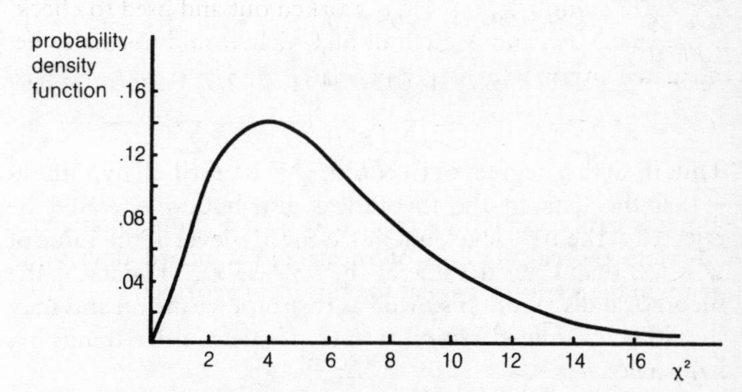

Fig. 12. **Chi-squared distribution.** Graph for n=7.

The χ^2 distribution is used in a number of statistical tests including the CHI-SQUARED TEST, FRIEDMAN'S TWO-WAY ANALYSIS OF VARIANCE BY RANK, and KRUSKAL-WALLIS ONE-WAY ANALYSIS OF VARIANCE.

chi-squared (or χ^2) test, *n.* a test of how well observed data fit a theoretical distribution. The data are collected into groups, and the number of observations in each group denoted by O (observed). The expected number in each group E is calculated from the theoretical distribution. The value of χ^2 is then worked out using

$$\chi^2 = \sum_{\substack{\text{All}\\ \text{groups}}} \frac{(O-E)^2}{E}$$

A small value of χ^2 means the data fit the distribution well, a large value badly.

The χ^2 test is actually a test of the NULL HYPOTHESIS that the data are a sample drawn from a parent distribution fitting the theoretical distribution being tested. CRITICAL VALUES of χ^2 for various SIGNIFICANCE LEVELS are given in tables (see p. 249), for different numbers of DEGREES OF FREEDOM ν as in the extract below for $\nu=5$.

Significance level %	99.5	99	97.5	95	10	5	2.5	1	0.5	.1
Critical value $\nu=5$	.412	.554	.831	1.15	9.24	11.07	12.83	15.09	16.75	20.52

Thus if, with 5 degrees of freedom, $\chi^2=10$, the null hypothesis – that the data fit the theoretical distribution – would be rejected at the 10% level, but not at the 5% level. If the value of χ^2 is less than critical value at the 95% level, the data fit the theoretical distribution so well as to arouse suspicion and they should be examined critically to see if one of three things has happened:

(a) The data have been invented;

(b) The theory has been formed round the data, and then the data are being used to confirm the same theory;

(c) Data which do not fit well have been rejected.

If none of these is the case, the data fit the theory well.

The number of degrees of freedom is calculated using

$$\text{Degrees of freedom} = \text{Groups} - \text{Restrictions}$$

Before the various values of E can be worked out, the observed data may have to be used to estimate certain parameters for the theoretical distributions, like its mean, μ, and standard deviation, σ. Each parameter that has to be estimated counts as one restriction. One degree of freedom is lost in addition, because once the numbers in $N-1$ groups are known, the number in the Nth group is also known. The table gives the number of degrees of freedom in some common cases.

Theoretical distribution	*Groups*	*Total sample size*	*Parameters needed*	*Total number of Restrictions*	*Degrees of freedom*
Normal	N	n	μ, σ	3	$N-3$
Binomial	N	n	p	2	$N-2$
Poisson	N	n	μ	2	$N-2$
Specified proportion	N	n		1	$N-1$
u x v Contingency Table	uv	n	Row and column totals	$(u-1)+(v-1)+1$	$(u-1)(v-1)$

The χ^2 test should generally only be applied to groups with expected size of at least 5. If a group is smaller than this, it should be combined with one of the other groups.

Example 1: a die is thrown repeatedly with results as follows:

Score	1	2	3	4	5	6
Frequency	3	7	8	10	14	18

Is there evidence, at the 5% significance level, to suggest that the die is biased?

Null hypothesis: The die is unbiased; all six scores were equally likely, each with a probability of 1/6.

The total number of throws is:

$$3+7+8+10+14+18=60$$

This is the only parameter needed to work out the values of E, based on the null hypothesis, to be $1/6 \times 60 = 10$ each. So there is only one restriction: $\nu = 6-1 = 5$. (The expected distribution is in a specified proportion, 1:1:1:1:1:1.)

Score	*O*	*E*	*O−E*	*(O−E)²/E*
1	3	10	−7	4.9
2	7	10	−3	0.9
3	8	10	−2	0.4
4	10	10	0	0
5	14	10	4	1.6
6	18	10	8	6.4

$$\sum \frac{(O-E)^2}{E} = 14.2$$

The critical value for χ^2 for $\nu = 5$ at the 5% level (see above) is 11.07. Since $14.2 > 11.07$, the null hypothesis is rejected. The evidence supports the idea that the die is biased.

Example 2: in a survey on class mobility, a research student interviewed 200 men, asking them about their jobs and those of their fathers. As a result he classified them, using his own scale, as follows:

O		*FATHER* *Upper*	*Middle*	*Lower*	
	Upper	16	25	19	60
SON	*Middle*	15	33	22	70
	Lower	19	22	29	70
Total		50	80	70	200

He wishes to test the null hypothesis that the class of a son is unrelated to that of his father, at the 10% level.
This theory gives expected values for E as follows:

E	*U*	*M*	*L*
U	$\frac{50}{200} \times 60$	$\frac{80}{200} \times 60$	$\frac{70}{200} \times 60$
M	$\frac{50}{200} \times 70$	$\frac{80}{200} \times 70$	$\frac{70}{200} \times 70$
L	$\frac{50}{200} \times 70$	$\frac{80}{200} \times 70$	$\frac{70}{200} \times 70$

E		*U*	*M*	*L*	
	U	15	24	21	60
=	*M*	17.5	28	24.5	70
	L	17.5	28	24.5	70
Total		50	80	70	200

So $\chi^2 = \sum \frac{(O-E)^2}{E}$

$$= \frac{1^2}{15} + \frac{1^2}{24} + \frac{(-2)^2}{21} + \frac{(-2.5)^2}{17.5} + \frac{5^2}{28} + \frac{(-2.5)^2}{24.5} + \frac{1.5^2}{17.5} + \frac{(-6)^2}{28} + \frac{4.5^2}{24.5}$$

$$= 4.04$$

The number of degrees of freedom (for a 3×3 contingency table) is $\nu = (3-1) \times (3-1) = 4$. For $\nu = 4$, the critical value for χ^2 at the 10% level is 7.78. As $4.04 < 7.78$, there is no reason to reject the null hypothesis. The figures fit his theory satisfactorily.

Example 3: the scores of the football teams in a league one week were as follows:

Goals	0	1	2	3	4	5
Frequency (teams)	28	42	19	6	3	2

Test the theory that these figures fit a POISSON DISTRIBUTION, at the 5% level.

Null hypothesis: The figures are a sample drawn from a Poisson distribution.

To work out the expected frequencies E, it is necessary to know the total number of teams and the mean number of goals per team.

$$\textit{Total number} = 28 + 42 + 19 + 6 + 3 + 2 = 100$$

$$Mean = \frac{\text{Total goals}}{\text{Number of teams}} = \frac{0 \times 28 + 1 \times 42 + 2 \times 19 + 3 \times 6 + 4 \times 3 + 5 \times 2}{100} = 1.2$$

There are thus two parameters found, and so two restrictions.
The values of E are given by the terms of

$$100e^{-1.2}\left(1 + 1.2 + \frac{1.2^2}{2!} + \frac{1.2^3}{3!} + \frac{1.2^4}{4!} + \frac{1.2^5}{5!} + \ldots\right)$$

	0	1	2	3	4	5	
or	30.12	36.14	21.69	8.67	2.60	.62	...

Since 2.60, .62, and subsequent numbers are all less than 5, a single group is formed for 3 goals or more.

Group	*O*	*E*	*O − E*	$(O-E)^2/E$
0	28	30.12	−2.12	.1492
1	42	36.14	5.86	.9502
2	19	21.69	−2.69	.3336
⩾3	11	12.05	−1.05	.0915
				$1.5245 = \chi^2$

There are 4 groups and 2 restrictions and so $\nu = 4 - 2 = 2$.
The critical value for $\nu = 2$ for the 5% level is 5.99. $1.5245 < 5.99$, so there is no reason to reject the null hypothesis. The figures fit a Poisson distribution satisfactorily.

chi-squared test for variance, *n.* a test of the NULL HYPOTHESIS, that a particular sample has been drawn from a parent population with a given VARIANCE. This test can only be applied if the parent population is normal. It depends upon the fact that the statistic

$$\frac{n\hat{\sigma}^2}{\sigma^2}$$

has a χ^2 distribution with $n - 1$ DEGREES OF FREEDOM, where n is

the sample size, $\hat{\sigma}$ is the parent standard deviation estimated from the sample and σ is the given parent standard deviation.
Example: test, at the 5% SIGNIFICANCE LEVEL, whether the sample 9, 5, 6, 4, 3, 5, 4, 6, 2 could have been drawn from a parent population with variance 2.
Null hypothesis: The parent variance $\sigma^2 = 2$.
Alternative hypothesis: $\sigma^2 \neq 2$
A two-tail test (see ONE- AND TWO-TAIL TESTS) is needed. From the sample data, sample variance is:

$$s^2 = \frac{\sum x^2}{n} - \left(\frac{\sum x}{n}\right)^2 = \frac{248}{9} - \left(\frac{44}{9}\right)^2 = 3.654$$

Estimated parent variance,

$$\hat{\sigma}^2 = s^2 \frac{n}{(n-1)} = 3.654 \times \frac{9}{8} = 4.111$$

The value of χ^2 obtained is given by

$$\chi^2 = \frac{n\hat{\sigma}^2}{\sigma^2} = \frac{9 \times 4.111}{2} = 18.50$$

The critical values of χ^2 are found from the tables on page 249. For $\nu = 9 - 1 = 8$ (see Fig. 13),

$$x_1 \text{ (the } 97\tfrac{1}{2}\% \text{ value)} = 2.18$$
$$x_2 \text{ (the } 2\tfrac{1}{2}\% \text{ value)} = 17.53$$

Since 18.50 does not lie between 2.18 and 17.53, the null hypothesis is rejected at this level.

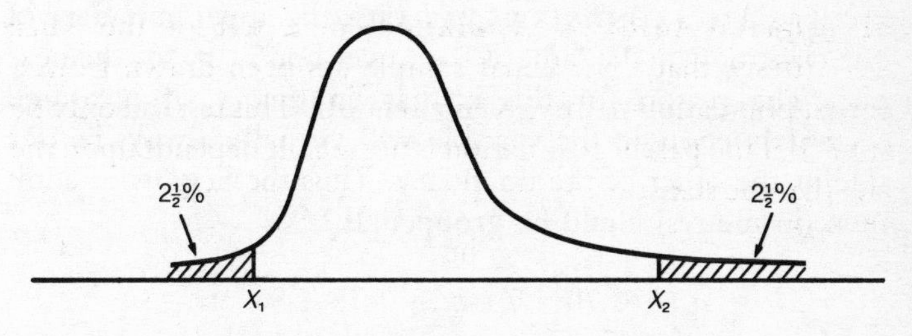

Fig. 13. **Chi-squared test for variance.**

Confidence limits for variance: if the same sample,

9, 5, 6, 4, 3, 5, 4, 6, 2

had been drawn from normal parent population with unknown variance, the parent variance could have been estimated as:

$$\hat{\sigma}^2 = s^2 \frac{n}{(n-1)} = 3.654 \times \frac{9}{8} = 4.111$$

95% confidence limits for an estimate are given by:

$$\frac{ns^2}{x_1} \quad \text{and} \quad \frac{ns^2}{x_2}$$

where x_1 and x_2 have the values already found, i.e. 2.18 and 17.53. These give values for the 95% confidence limits as:

$$\frac{9 \times 3.654}{2.18} \quad \text{and} \quad \frac{9 \times 3.654}{17.53}$$

$$15.09 \quad \text{and} \quad 1.88$$

When the variances of two samples are used to test if those samples could have been drawn from a common parent population (or populations with the same variance), the F-TEST is used.

class, *n*. a collection or division of people, things, or values of a variable, sharing a common characteristic. See GROUPED DATA. See also CLASS BOUNDARIES.

class boundaries, *n*. the boundaries for the classes when data is grouped. For a CONTINUOUS VARIABLE, the upper boundary of one class may be the same as the lower boundary for the next one. The reason for this is that, in theory, if measured accurately enough, the variable will virtually always be one side or the other of the boundary. Thus the heights of adult men (in metres) could be grouped as:

1.65–1.70 1.70–1.75 1.75–1.80 etc.

In practice, however, measurements are always rounded, in

some way or other, and so it is often convenient to make it clear which class a variable apparently on the boundary is to be assigned to. In that case the classes might be described as:

1.65– 1.70– 1.75– etc.
or $1.65 \leqslant \; < 1.70$ $\quad 1.70 \leqslant \; < 1.75$ $\quad 1.75 \leqslant \; < 1.80$

For a DISCRETE VARIABLE, on the other hand, class boundaries cannot be shared. The distribution of the annual egg yield per hen of a farmer's chickens might, for example, be grouped as

No. of eggs	0–49	50–99	100–149	150–199	200–249	250–299	300–349	Over 350
Frequency (no. of hens)	1	15	35	46	53	42	18	0

In this case the same figures do not appear in the boundaries of two classes. There is no need for this, since the number of eggs laid must be a whole number.

class interval, *n.* the interval between CLASS BOUNDARIES for grouped data, or the length of this interval. Symbol: c.

climograph, *n.* a graph showing rainfall against temperature for a particular place. Points are plotted for each month of the year, (MEAN values being taken), and then joined up to form a continuous 12-sided figure (see Fig. 14).

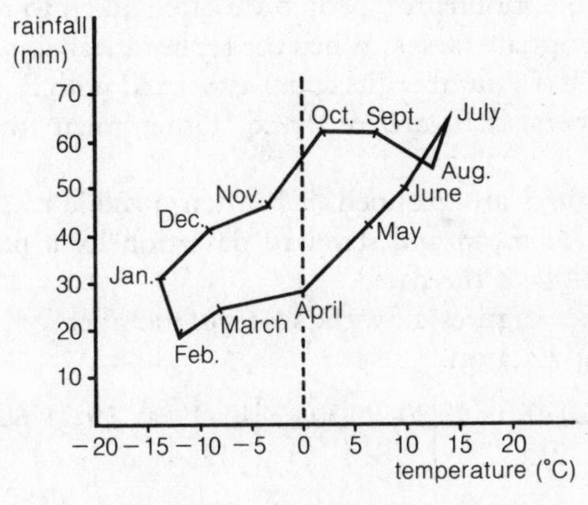

Fig. 14. **Climograph.** Rainfall for Arkhangelsk (USSR). Source: *Philip's Certificate Atlas*, 1978.

cluster sampling, *n.* a method of sampling in which the members of the sample are chosen from one or several groups (or clusters) rather than at random. From a statistical point of view this is less satisfactory than SIMPLE RANDOM SAMPLING, but it may well be more practical and/or economical.

Imagine a veterinary research officer investigating the incidence of various parasites among goats in Nigeria. To select, say, 100 random goats in the country, and travel backwards and forwards to see each one, would be very time-consuming and expensive; indeed, in the absence of some form of register of goats, it would also be very hard to achieve random selection. What he would do instead would be to inspect several goats at each of a number of villages selected at random from a register of villages. This would be cluster sampling. He might even take all 100 from the same location, which would be unsatisfactory, as it could, for instance, be the case that all the goats in any village are infected with the same parasites.

If the members of each cluster are somewhat homogeneous the effective sample size may be little more than the number of clusters.

coded data, *n.* data which has been translated from the form in which it was collected to one that is more convenient for analysis. In questionnaires, people are often asked to reply by ticking appropriate boxes. When the replies are analysed, each box usually has a number (its code) associated with it, and it is these numbers that are analysed rather than the ticks themselves.

When figures are grouped, it is often possible to simplify calculations of mean and standard deviation by a particular kind of coding of the data.

Example: these figures show the scores of members of a cricket team during a season.

Score (runs)	0–9	10–19	20–29	30–39	40–49	50–59	60–69	70–79
Frequency	21	24	18	13	12	6	4	2

The mid-interval values are $4\frac{1}{2}$, $14\frac{1}{2}$, . . . $74\frac{1}{2}$, which are not the easiest of numbers to deal with. The calculation of MEAN and

STANDARD DEVIATION can thus be simplified in this case by subtracting $4\frac{1}{2}$, and dividing by 10 to give the mid points of the groups coded values of 0, 1, 2, . . . 7, as is shown in the calculation below.

Interval	*Mid-interval point, X*	$X-4\frac{1}{2}$	$(X-4\frac{1}{2})/10$	*Frequency f*	$\left(\frac{X-4\frac{1}{2}}{10}\right)f$	$\left(\frac{X-4\frac{1}{2}}{10}\right)^2$	$\left(\frac{X-4\frac{1}{2}}{10}\right)^2 f$
0–9	$4\frac{1}{2}$	0	0	21	0	0	0
10–19	$14\frac{1}{2}$	10	1	24	24	1	24
20–29	$24\frac{1}{2}$	20	2	18	36	4	72
30–39	$34\frac{1}{2}$	30	3	13	39	9	117
40–49	$44\frac{1}{2}$	40	4	12	48	16	192
50–59	$54\frac{1}{2}$	50	5	6	30	25	150
60–69	$64\frac{1}{2}$	60	6	4	24	36	144
70–79	$74\frac{1}{2}$	70	7	2	14	49	98
				100	215		797

For the coded data, the mean is $215 \div 100 = 2.15$, and the standard deviation is

$$\sqrt{\frac{797}{100} - 2.15^2} = 1.83$$

These values can then be uncoded, to give

$$Mean = 2.15 \times 10 + 4.5 = 26$$
$$Standard\ deviation = 1.83 \times 10 = 18.3$$

The coding could also have been done by subtracting, say, $34\frac{1}{2}$ from each mid-interval point, and then dividing by 10, which would have given coded values -3, -2, -1, 0, 1, 2, 3 and 4. This would have reduced the size of the numbers in the calculation still further.

coefficient, *n.* **1.** an index of measurement of a characteristic. Examples are the COEFFICIENT OF ALIENATION, and the COEFFICIENT OF DETERMINATION. **2.** a numerical or constant factor in an algebraic expression. For example, 5 and 7 are the coefficients of x^2 and x^3 respectively in the expression $5x^2 + 7x^3$.

coefficient of alienation, *n.* a measure of the extent to which two random variables are unrelated. The coefficient of alienation is given by:

$$\sqrt{(1-r^2)}$$

where r is the product-moment CORRELATION COEFFICIENT.

coefficient of determination, *n.* the proportion of the variation of the dependent variable which is taken up by fitting it to the REGRESSION LINE. The variation in the set $y_1, y_2, \ldots, y_n$, with mean $\bar{y}$, is given by:

$$\sum_{r=1}^{n} (y_r - \bar{y})^2$$

This can be split into two parts:

$$\sum_{r=1}^{n} (y_r - \bar{y})^2 = \underbrace{\sum_{r=1}^{n} (y_r - y')^2}_{\text{Unexplained variation}} + \underbrace{\sum_{r=1}^{n} (y'_r - \bar{y})^2}_{\text{Explained variation}}$$

where (x_r, y'_r) is the point on the y on x regression line vertically above or below the point (x_r, y_r). Thus the unexplained variation is the sum of the squares of the RESIDUALS $d_1, d_2, d_3, \ldots$.

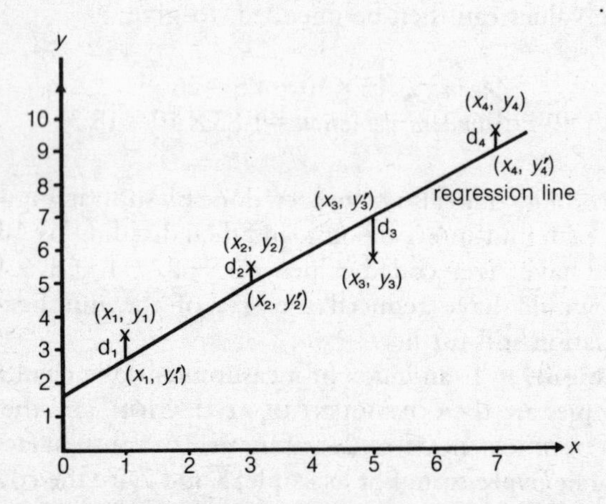

Fig. 15. **Coefficient of determination.** Graph showing the residuals d_1, d_2, d_3 and d_4 for the set of points $(x_1, y_1) \ldots (x_4, y_4)$.

The coefficient of determination is thus Explained variation/Total variation, and has the value r^2, where r is the product-moment CORRELATION COEFFICIENT. These statements are also true if the word, VARIANCE, is substituted for variation so that

$$\textit{Coefficient of determination} = \frac{\text{Explained variance}}{\text{Total variance}} = r^2$$

Example: find the coefficient of determination for the points (1, 3), (3, 5), (5, 6) and (7, 10), plotted on the graph (Fig. 15).

x	y	$x-\bar{x}$	$y-\bar{y}$	$(x-\bar{x})^2$	$(y-\bar{y})^2$	$(x-\bar{x})(y-\bar{y})$
1	3	−3	−3	9	9	9
3	5	−1	−1	1	1	1
5	6	1	0	1	0	0
7	10	3	4	9	16	12
16	24			20	26	22

$\bar{x}=4 \quad \bar{y}=6 \quad s_x^2=5 \quad s_y^2=6.5 \quad s_{xy}=5.5$

From these figures, the regression line can be calculated:

$$y-\bar{y}=\frac{s_{xy}}{s_x^2}(x-\bar{x}) \qquad y-6=\frac{5.5}{5}(x-4)$$

which simplifies to $y=1.1x+1.6$.
So

x	y	$y'=1.1x+1.6$	$d=(y-y')$	d^2
1	3	2.7	.3	.09
3	5	4.9	.1	.01
5	6	7.1	−1.1	1.21
7	10	9.3	.7	.49
16	24			1.8

Total variation $\sum(y-\bar{y})^2=26$
Unexplained variation $\sum d^2=1.8$
Explained variation $=24.2$

Coefficient of determination

$$= \frac{\text{Explained variation}}{\text{Total variation}} = \frac{24.2}{26} =$$
$$= .931$$

From this it follows that $r^2 = .931$.
Product-moment correlation coefficient $r = .965$.
This value for the correlation coefficient is the same as that calculated from the formula

$$r = \frac{s_{xy}}{s_x s_y} = \frac{5.5}{\sqrt{5}\sqrt{6.5}} = .965$$

coefficient of kurtosis, see KURTOSIS.
coefficient of mean deviation, *n.* an ABSOLUTE MEASURE OF DISPERSION.
coefficient of skewness, see SKEW.
coefficient of variation, *n.* an ABSOLUTE MEASURE OF DISPERSION, sometimes given as a percentage.

$$\textit{Coefficient of variation} = \frac{\text{standard deviation}}{\text{mean}}$$

combinations, *n.* the ways of selecting a subset of a set, where the order in which the elements are selected is of no importance. Thus the possible combinations of three letters out of A, B, C, D and E are 10 in all:

A B C	A B D	A B E
A C D	A C E	A D E
B C D	B C E	B D E
C D E		

The number of ways of selecting r objects from a total of n, all different, is denoted by:

$$^{n}C_{r} = \frac{n!}{r!(n-r)!}$$

$^{n}C_{r}$ also gives the coefficient of $x^{n-r}y^{r}$ in the binomial expansion of $(x+y)^{n}$ (see also FACTORIAL).

complementary events, *n.* events which are both EXCLUSIVE and EXHAUSTIVE. The events, 'The television is turned on' and 'The television is turned off,' are complementary. They cannot both happen at the same time (they are exclusive), and they cover all possibilities (they are exhaustive).

complementary sets, *n.* DISJOINT sets whose UNION is the UNIVERSAL SET $\mathscr{E}$.

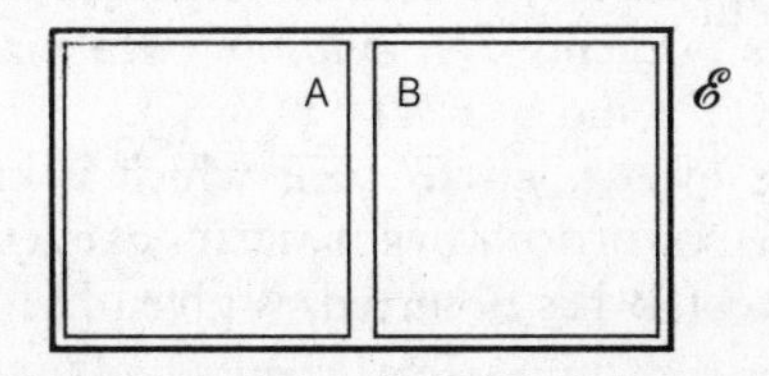

Fig. 16. **Complementary sets.** Venn diagram showing that A and B are complementary sets.

In the VENN DIAGRAM (Fig. 16), $A \cap B = \emptyset$ and $A \cup B = \mathscr{E}$, so that the sets A and B are complementary.

The complement of a set X is denoted by X′, the set of all elements belonging to the universal set, but not included in X; X′ may be thought of as the set, not X. X and X′ are complementary sets. Thus, $X \cap X' = \emptyset$, and $X \cup X' = \mathscr{E}$.

completely randomized design, see EXPERIMENTAL DESIGN.

concordance, see KENDALL'S COEFFICIENT OF CONCORDANCE.

conditional probability, *n.* the probability p of an event occurring, given that another event has already occurred. The usual notation for the combined event of 'Event A occurring given that event B has occurred' is A|B (A given B). The probability is given by:

$$p(A|B) = \frac{p(A \cap B)}{p(B)}$$

Example: a card player cheats by looking at one of his opponent's cards. He sees a black picture card (King, Queen, or Jack) but he cannot make out which one. What is the probability that the card is a King?

Event A: The card is a King.
Event B: The card is a black picture card.

$$p(B)=\frac{6}{52} \quad p(A)=\frac{4}{52} \quad p(A \cap B)=\frac{2}{52}$$

$$p(A|B)=\frac{p(A \cap B)}{p(B)}=\frac{2/52}{6/52}=\frac{1}{3}$$

Conditional probability is used in the definition of INDEPENDENT EVENTS. Events X and Y are independent if $p(X|Y)=p(X|Y')$ and $p(Y|X)=p(Y|X')$.

conditioning event, *n.* an event which is known to have occurred. The CONDITIONAL PROBABILITY of event A occurring, given that event B has occurred, is given by:

$$p(A|B)=\frac{p(A \cap B)}{p(B)}$$

In this, B is the conditioning event.

confidence interval, *n.* that interval within which a PARAMETER of a PARENT POPULATION is calculated (on the basis of sample data) to have a stated probability of lying. The larger the sample size n, the smaller is the confidence interval; in other words, the more accurate is the estimate of the parent mean. Example: a sample of 20 examination papers have the following marks (%);

22,	26,	30,	34,	47,	48,	50,	50,	52,	53
53,	53,	55,	57,	57,	61,	64,	72,	74,	82

The mean of these figures is 52, and so it is estimated that the mean for the examination as a whole is 52%, i.e., it is the parent mean. It is, however, very unlikely that the true parent mean is exactly 52%, although it will probably be somewhere near that figure. The question which therefore arises is, 'How accurate is the figure of 52%?'

The 95% confidence interval is the interval produced in such a way that there is a probability of 95% that the true parent mean lies inside it. So, if a large number of samples of 20 papers were taken from the same examination, and if a 95% confidence interval for the parent mean were worked out from each sample, then 95% of the intervals would include the true parent mean; the other 5% would not. The upper and lower bounds of the 95% confidence interval are the 95% CONFIDENCE LIMITS for the parent mean. Confidence limits may be taken for other levels, e.g., 90%, 99%, 99.9%, etc. Confidence limits for the mean of a distribution are worked out using the formula:

$$\text{sample mean} \pm k \times \frac{\text{parent standard deviation}}{\sqrt{\text{n}}}$$

where k is a number whose value depends on the confidence level, and the circumstances of the sampling. Provided that either the sample is large, or the sample is from a normal population whose standard deviation is known, k can be worked out from NORMAL DISTRIBUTION tables (p. 247).

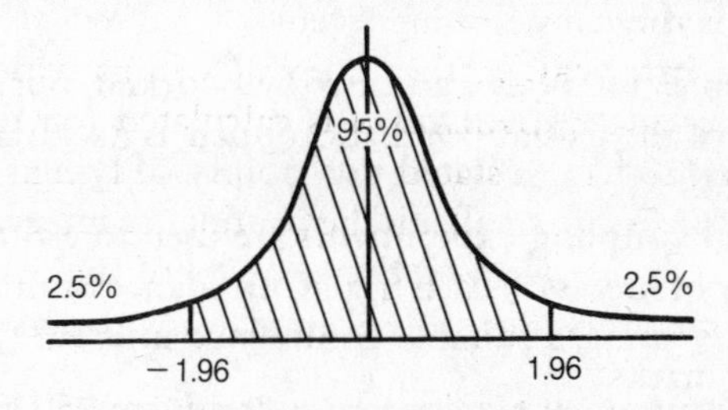

Fig. 17. **Confidence interval.**

For 95% limits, the two unshaded areas in Fig. 17 are both 2.5%, corresponding to ± 1.96 standard deviations, where

$$k=1.645 \quad \text{for 90\% limits}$$
$$k=1.96 \quad \text{for 95\% limits}$$
$$k=2.58 \quad \text{for 99\% limits}$$

In the example of the examination marks, n is not large (20), and the parent standard deviation is unknown. In that situation k has to be calculated on the basis of the t-DISTRIBUTION rather than the normal distribution. The sample size n is 20 and so ν, the number of DEGREES OF FREEDOM, is $20-1=19$ (one degree of freedom is lost by estimating the mean from the sample). For $\nu=19$, and 95% limits, the table of critical values of t (see p. 248) gives $k=2.093$.

The parent standard deviation is estimated using:

$$\hat{\sigma}=s\sqrt{\frac{n}{n-1}}=15.01\sqrt{\frac{20}{19}}=15.40$$

where $\hat{\sigma}$ is the estimated parent standard deviation and s is the sample standard deviation.

So the 95% confidence limits for the parent mean are:

$$\text{Sample mean} \pm k \times \frac{\text{estimated parent SD}}{\sqrt{n}} = 52 \pm 2.093 \times \frac{15.40}{\sqrt{20}}$$

$$= 44.8 \text{ and } 59.2$$

The 95% confidence interval for the parent mean is 44.8 to 59.2.

Confidence intervals can also be worked out for other properties of the parent population, such as its variance, from sample data.

Binomial sampling experiments are used to estimate p, the probability of success in each trial. Confidence limits for p may be found by using a suitable confidence interval chart.

Example: a binomial experiment is conducted 50 times, with 32 successes. The probability of success p is estimated as:

$$\hat{p}=\frac{32}{50}=.64$$

95% confidence intervals for p are found from a chart for binomial sampling experiments (Fig. 18) by locating $\hat{p}=.64$ on the horizontal axis; finding the intersections of the vertical line through .64 with the two curves marked 50 (the number of trials); then reading the equivalent values of p (.49 and .77) on the vertical scale. Thus, in this example, the 95% confidence limits are .49 and .77.

For the use of confidence interval charts for population correlation coefficients, see FISHER'S Z-TRANSFORMATION.

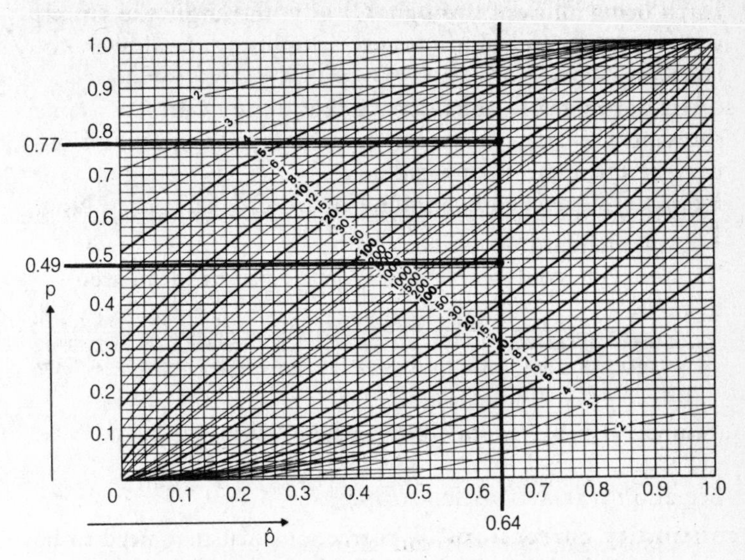

Fig. 18. **Confidence interval.** 95% confidence interval chart for binomial sampling experiments. Source: *Elementary Statistics Tables*, H.R. Neave, George Allen and Unwin.

confidence limits, *n.* the lower and upper boundaries of a CONFIDENCE INTERVAL.

consistent, *adj.* (of an ESTIMATOR) such that the larger the sample size, the more accurate is the estimate. The data are used to form an estimate $\hat{\theta}_n$ of the value θ of a parameter of a parent population from a sample of size n. Often, but not invariably, the parameter estimated is the mean or standard deviation.

An estimate is said to be consistent if, the larger the sample size n, the closer the estimate $\hat{\theta}_n$ is to the true value θ. This can be stated more formally as:

Probability $(|\hat{\theta}_n - \theta| < \epsilon) \rightarrow 1$ as $n \rightarrow \infty$ for all $\epsilon > 0$.

constant, *n.* a quantity, or its algebraic representation, whose value remains invariable.

consumer's risk, *n.* (in quality control) the probability of a consignment with the proportion of defectives equal to the LOT

TOLERANCE PERCENTAGE DEFECTIVE being accepted, despite the inspection scheme. This is thus the probability of a borderline batch being allowed through. Clearly, the larger the sample size (meaning more thorough sampling), the lower the consumer's risk.

contingency table or **bivariate frequency table,** *n.* a table showing a bivariate frequency distribution. The classifications may be quantitative or qualitative.
Example: Associated Examining Board O-level entries, Nov. 1982.

		AGE OF CANDIDATE							
		<15	*15*	*16*	*17*	*18*	*19*	*20*	*>20*
SEX OF CANDIDATE	*Female*	100	6,322	13,449	9,387	2,969	1,045	592	2,812
	Male	164	6,077	12,212	7,909	2,779	1,285	824	3,176

(Source: *A.E.B. Statistics*, November 1980/81/82.)

See also BIVARIATE DISTRIBUTION.

continuity corrections, *n.* corrections that may need to be applied when the distribution of a discrete random variable is approximated by that of a continuous random variable.
Example: Intelligence Quotient, as measured by certain tests, is an integer. Its distribution has mean 100 and standard deviation 15. It is thus a discrete variable. Its distribution may however be approximated by the NORMAL DISTRIBUTION, with the same mean and standard deviation. This in effect involves fitting the normal curve to a histogram (Fig. 19a). When this is done, the space that meant 100 on the histogram takes on the meaning 99.5–100.5 for the normal distribution curve.

The question, 'What proportion of people have an IQ between 110 and 115 (inclusive)?', could be answered by giving the areas of the bars labelled 110 to 115. If, however, the normal distribution is used, the correct interval is 109.5 to 115.5 This correction of 0.5 at either end of the required interval is called a continuity correction (Fig. 19b). The area is then:

$$\Phi\left(\frac{115.5-100}{15}\right)-\Phi\left(\frac{109.5-100}{15}\right)=.113$$

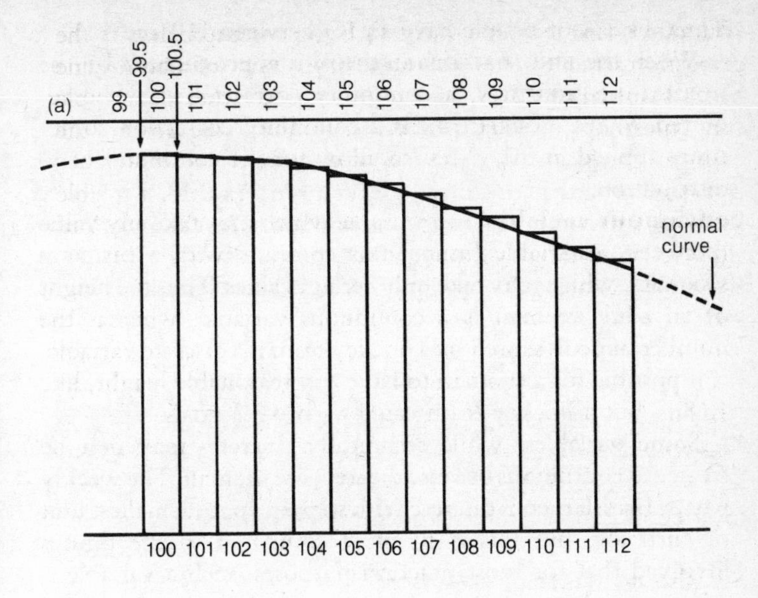

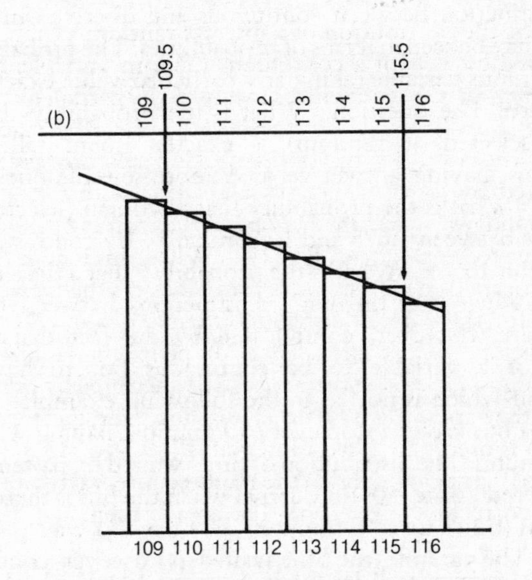

Fig. 19. **Continuity corrections.** (a) and (b).

Thus, 11.3% of people have an IQ between 110 and 115.

When the BINOMIAL DISTRIBUTION is approximated by the NORMAL DISTRIBUTION, a continuity correction may be required. YATES' CORRECTION is a continuity correction sometimes applied in the χ^2-test to allow for the continuity of χ^2 distribution.

continuous variable, *n.* a variable which may take any value (between reasonable limits); this contrasts with a DISCRETE VARIABLE, which may take only certain values. Thus the height of an adult woman is a continuous variable, whereas the number of goals scored by a hockey team is a discrete variable. It is possible for a woman to have any reasonable height, like 1.68m, but a hockey team cannot score 2.3 goals.

Some variables, while technically discrete, may best be treated as continuous because the steps are so small. The weekly pay of British men is discrete with steps of 1p, the smallest unit of currency, but this is so small compared to the figures involved that for most practical purposes, such a variable is continuous.

The distinction between continuous and discrete random variables may be seen in terms of probabilities. The probability of a continuous variable taking any particular value exactly is usually zero. The question, 'What is the probability that a woman (selected at random) is exactly 1.68m tall?', is meaningless, having answer zero. A more sensible question would be, 'What is the probability that a woman (selected at random) is between 1.675 and 1.685m tall?'. By contrast, it is quite sensible to ask, 'What is the probability that a die shows exactly 5 when it is thrown?'. Distinctions between these variables are, however, complicated by the fact that it is possible for a variable to be continuous but to have a distribution which is not, as in the following example.

Example: a bus leaves a stop every 10 minutes, having waited there 2 minutes. The distribution of times waited by passengers can be discrete (time = 0, they arrive when the bus is there) or continuous (times up to 8 minutes; they have to wait for the next bus). The variable, the time waited, is however, continuous. If the distribution of a continuous variable is not continuous, then the probability of it taking a particular exact value need not be zero, as in the case of time = 0 when waiting for the bus.

control charts, *n.* (in quality control) diagrams used to give warning of when a machine or process is not functioning well. In this context manufacturing processes may be divided into two types. There are those whose product is either good or faulty, like electric light bulbs, and others where the quality of the product is determined by measurement. If a machine for making 5cm nails is producing them with a mean length of 4.8cm, such nails are unsatisfactory, as they do not meet the specification; they can, however, still be used as nails, whereas faulty light bulbs are quite worthless.

In the situation where the product is either good or faulty, an inspection scheme would involve recording the numbers of

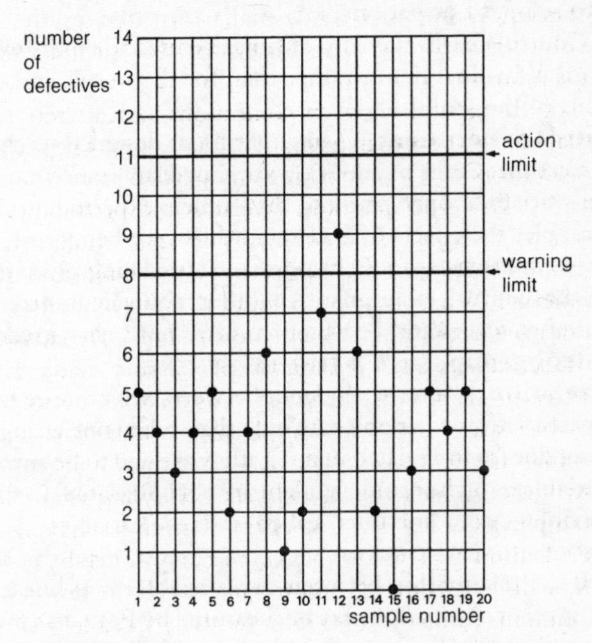

Fig. 20. **Control charts.** The chart for numbers of defectives in samples of size 100 (PROCESS AVERAGE, 4%).

defectives in samples of a certain size. This information could then be plotted on a control chart. This chart would have two lines drawn on it, the inner control (or warning) limit and the outer control (or action) limit; see Fig. 20. The warning limit is usually set so that if the process is running correctly it is exceeded by 5% of the samples, the action limit by 0.1%.

In cases where a measurement is involved, two things need to be watched, the mean of the variable and its dispersion. A change in either could result in an unduly high proportion of the products being outside acceptable tolerance limits. Two control charts are thus needed, one for the mean, the other for the standard deviation of the samples. The control limits are typically set so that 5% of items from a correctly functioning machine lie outside the warning limits, 0.1% outside the action limits. If the standard deviation is not known it is often estimated using the tables on page 257, for converting RANGE into standard deviation.

Control charts may also at times be used for measurements from individual items, rather than for the means and dispersions of the samples.

control experiment, *n.* an experiment designed to check or correct the results of another experiment by removing one of the variables operating in that other experiment. If, for example, the effects of a sleeping pill were being tested, one group of subjects would be observed after being given the pill. In the control experiment, another equivalent group, the *control group*, would be observed after not being given it.

control group, see CONTROL EXPERIMENT.

correlation, *n.* interdependence between two or more random variables. If two variables are such that, when one changes, the other does so in a related manner, they are said to be correlated. Variables which are independent are not correlated. Correlated variables are sometimes called related variables. See also ASSOCIATION, CAUSATION.

If the relationship between two variables is of linear form, the level of correlation may be measured by Pearson's product-moment correlation coefficient. If the relationship is such that increasing the size of one variable changes the size of the other in a fixed direction (either increasing or decreasing), then Kendall's coefficient of rank correlation or Spearman's rank correlation coefficient may be used as a measure of the level of correlation (see CORRELATION COEFFICIENT).

It is possible for variables to be correlated but for none of these measures to be appropriate, as in the case illustrated in Fig. 21.

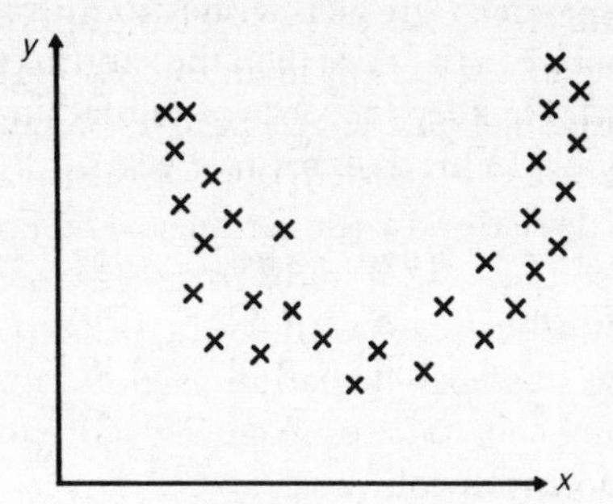

Fig. 21. **Non-linear correlation.**

correlation coefficient, *n.* a measure of ASSOCIATION between two random variables. If two random variables are such that when one changes the other does so in a related manner, they are said to be correlated (see CORRELATION). Variables which are independent are not correlated. The correlation coefficient is a number between −1 and 1. It is calculated from a number of pairs of observations which are normally referred to as points (x, y). A coefficient of 1 means perfect positive correlation, −1 perfect negative correlation, and 0 no correlation. See also ALIENATION.

There are several ways of calculating correlation coefficients.

(a) *Pearson's product moment correlation coefficient* (symbol usually r). This is a measure of linear association and so is unsuitable if the relationship is non-linear.

This is calculated from the formula

$$\frac{s_{xy}}{s_x s_y}$$

where s_x and s_y are the standard deviations of x and y, and s_{xy} is the covariance of x and y, given by:

$$s_{xy} = \sum_{i=1}^{i=n} \frac{x_i y_i}{n} - \bar{x}\bar{y} \quad \text{or} \quad \sum_{i=1}^{n} \frac{(x_i - \bar{x})(y_i - \bar{y})}{n}$$

Example: a spaghetti grower wishes to know if there is any correlation between the rainfall in the month of March and the ultimate yield. He keeps records of rainfall in this month and the average yield of his trees for five years.

	1979	*1980*	*1981*	*1982*	*1983*
x (Rainfall cm)	28	32	29	41	30
y (Yield kg)	134	142	136	168	150

The calculation is as follows:

	x	y	$x-\bar{x}$	$y-\bar{y}$	$(x-\bar{x})^2$	$(y-\bar{y})^2$	$(x-\bar{x})(y-\bar{y})$
	28	134	−4	−12	16	144	48
	32	142	0	−4	0	16	0
	29	136	−3	−10	9	100	30
	41	168	9	22	81	484	198
	30	150	−2	4	4	16	−8
÷n	5)160	5)730			5)110	5)760	5)268
	$\bar{x}=32$	$\bar{y}=146$			$s_x^2=22$	$s_y^2=152$	$s_{xy}=53.6$

Thus $\bar{x}=32$, $\bar{y}=146$, $s_x=\sqrt{22}$, $s_y=\sqrt{152}$ and $s_{xy}=53.6$; the correlation coefficient r is:

$$r = \frac{53.6}{\sqrt{22}\,\sqrt{152}} = .927$$

(b) *Rank correlation coefficients*. There are occasions when it is not convenient, economic, or even possible, to give values to variables, only rank orders. In such cases, a rank correlation coefficient has to be used. Rank correlation coefficients may also be appropriate in cases of non-linear relationship between the variables.

Kendall's coefficient of rank correlation: symbol: τ. This is calculated for n pairs of values using the formula:

$$\tau = \frac{S}{\frac{1}{2}n(n-1)}$$

where S is a sum of scores; its computation is explained in the example below.
Example: in a beauty competition there are two judges, O'Hara and O'Shaughnessy. They place the girls as follows:

	Annie	*Bridget*	*Charmain*	*Dolly*	*Ela*
O'Hara	1	5	2	4	3
O'Shaughnessy	2	4	1	3	5

What is the rank correlation coefficient between their judgements?

To calculate this, one of the sets of results is put in order, and the other is placed beneath it. The score S for each girl is then worked as follows:

	Annie		*Charmain*		*Ela*		*Dolly*	*Bridget*
O'Hara	1		2		3		4	5
O'Shaughnessy	2		1		5		3	4
Score S=	2	+	3	+	−2	+	1	−

Total score=4

$$\textit{Maximum possible score} = \frac{n(n-1)}{2} \text{ or } \tfrac{1}{2}n(n-1) = 10$$

$$\textit{Kendall's rank correlation coefficient } T = \frac{4}{10} = .4$$

Each entry in the second row (i.e. from O'Shaughnessy) is compared in turn with each of the entries to the right of it. If the entry to the right is the greater +1 is counted, if it is less −1, and if it is equal, 0.

Thus the entry under Annie (2) is scored

2	1	5	3	4	
	−1	+1	+1	+1	=2

The total score (4) is the sum of the scores for each girl, in this case,

$$2 \quad + \quad 3 \quad + \quad -2 \quad + \quad 1 \quad = 4$$

The last entry (i.e. Bridget), having no numbers to the right of it, makes no contribution to the total score.

The bottom line of the formula

$$T = \frac{S}{\frac{1}{2}n(n-1)}$$

i.e. $\frac{1}{2}n(n-1)$, represents the maximum possible score which would occur if there were perfect agreement (10). In this case Kendall's rank correlation coefficient T is $\frac{4}{10} = .4$.

Spearman's coefficient of rank correlation: symbol, ρ. This is calculated by the formula:

$$1 - \frac{6\Sigma D^2}{n(n^2-1)}$$

where D is the rank difference for any member.

Example: taking the same data for the beauty competition used in the previous section, the calculation is as follows:

	Annie	*Bridget*	*Charmain*	*Dolly*	*Ela*
O'Hara	1	5	2	4	3
O'Shaughnessy	2	4	1	3	5
D	−1	1	1	1	−2
D^2	1	1	1	1	4

The value of D is obtained simply by subtracting O'Shaughnessy's rank from that of O'Hara, for each girl; the square of each value, D^2, is then obtained.

Thus, $\Sigma D^2 = 1+1+1+1+4 = 8$, and $n = 5$.

Spearman's rank correlation coefficient is then obtained, as

$$1 - \frac{6\Sigma D^2}{n(n^2-1)} = 1 - \frac{6 \times 8}{5(25-1)} = 0.6$$

Interpretation of correlation coefficients. Sometimes a correlation coefficient is worked out because its value is of interest, at others as a test of possible correlation between the variables.

Test of possible correlation. The NULL HYPOTHESIS for this test is that the points taken are a sample from a parent population in which the two variables are uncorrelated. The figure obtained is then interpreted in terms of its SIGNIFICANCE LEVEL.
Example: using the example of the spaghetti trees, there are five pairs of results, and the (Pearson's product-moment) correlation coefficient is 0.927. The number of DEGREES OF FREEDOM ν is two less than the number of points, $\nu = n - 2$, and so in this case $\nu = 5 - 2 = 3$.
The tables on page 254 give, for $\nu = 3$,

ν	.10	.05	.02	.01
3	.805	.878	.934	.959

Thus, .927 lies between .05 and .02, or 5% and 2%. This is the significance level, the probability that, if the null hypothesis of no underlying correlation is true, a correlation coefficient as large (or larger) than that found could have arisen from a random sample. These figures may be interpreted as follows:

10%	possible correlation
5%	probable correlation
1%	very probable correlation
.1%	almost certain correlation

The three different correlation coefficients given in this entry, r, τ and ρ, all have their own tables of significance (see pp. 254, 255, 256). For the use of a confidence-interval chart for r, see Fig. 41 (FISHER'S Z-TRANSFORMATION).
Test for non-zero correlation coefficient. To test the null hypothesis that the parent population has a correlation coefficient other than zero, Fisher's z-transformation is used. This also allows CONFIDENCE LIMITS to be set up for an estimated value of the parent population's correlation coefficient.

countable set, *n.* a non-empty SET, each of whose members can be labelled with a distinct positive integer. If the elements of such a set can be labelled using the integers 1, 2, . . . N, the set is said to be a *finite set,* with N elements. If no such N exists, the set is said to be *countably infinite.*

The set of prime factors of the number 462 is finite, with 4 elements (2, 3, 7 and 11).
The set of all prime numbers is countably infinite.
The set of all points on a plane is infinite, but not countably infinite.
The set of all points on a plane is infinite, but uncountable.

covariance, *n.* a statistic used for bivariate samples, symbol: s_{xy}, or a parameter of a bivariate distribution, symbol: cov (X, Y). Sample covariance is given by:

$$s_{xy} = \frac{\sum_{i=1}^{n} x_i y_i}{n} - \bar{x}\bar{y} = \sum_{i=1}^{n} \frac{(x_i - \bar{x})(y_i - \bar{y})}{n}$$

Sample covariance is used in working out the equations for REGRESSION LINES and the product-moment CORRELATION COEFFICIENT.

The covariance of two random variables, X and Y, is

$$\text{cov}(X, Y) = \text{E}(XY) - \text{E}(X)\text{E}(Y)$$

where E( ) denotes expected value, or expectation. For independent variables the covariance is zero. For a sample from a bivariate distribution in which the variables are independent, the expected value of the sample covariance is also zero.

Covariance appears in the relationship:

$$\text{Var}(A \pm B) = \text{Var}(A) + \text{Var}(B) \pm 2\,\text{cov}(A, B)$$

which reduces to:

$$\text{Var}(A \pm B) = \text{Var}(A) + \text{Var}(B)$$

if A and B are independent random variables.

Cramér-Rao lower bound, *n.* an expression for the smallest possible variance for an unbiased ESTIMATOR.

critical region, *n.* the region which leads to the rejection of the NULL HYPOTHESIS in a statistical test. This may take the following forms:

$\chi^2 >$. . .(CHI-SQUARED TEST)
$z >$. . .(NORMAL DISTRIBUTION)
$t >$. . .(t-TEST)
$r >$. . .(test for CORRELATION)

The critical region will depend on the SIGNIFICANCE LEVEL at which the test is applied. Thus, for the normal distribution two-tail test, the critical regions for z are:

$z > 1.645$ 10% level of significance
$z > 1.96$ 5% level of significance
$z > 2.58$ 1% level of significance

The values 1.645, 1.96 and 2.58, are called the CRITICAL VALUES for their respective significance levels.

critical value, *n.* the value of a test STATISTIC (like z, t, χ^2, etc.) which is the borderline between accepting and rejecting the NULL HYPOTHESIS. The critical value for any test depends on the significance level at which the test is being applied, and on whether it is a ONE- OR TWO-TAIL TEST.

cross-section analysis, *n.* statistical analysis of variations in data obtained from a set of sources at the same time. Compare TIME SERIES ANALYSIS.

cumulative distribution function, see DISTRIBUTION FUNCTION.

cumulative frequency, *n.* the frequency with which a variable has value less than or equal to a particular value. Example: the yield of each tree in an orchard of 200 trees is recorded, and the figures are grouped as follows.

Yield (kg)	0–10	10–20	20–30	30–40	40–50	50–60	60–70	70–80	80–90
Frequency (trees)	0	4	16	32	46	48	34	12	8

The corresponding cumulative frequency table is:

Yield	$\leqslant 10$	$\leqslant 20$	$\leqslant 30$	$\leqslant 40$	$\leqslant 50$	$\leqslant 60$	$\leqslant 70$	$\leqslant 80$	$\leqslant 90$
Frequency	0	4	20	52	98	146	180	192	200

This is derived from the frequency table by adding the frequencies from the left, or cumulating up to the entry in question. This can be drawn as a cumulative frequency graph or *ogive* (Fig. 22). Such a graph allows the median, quartiles, percentiles, etc., to be read easily. It also allows questions like, 'Estimate how many trees yielded less than 43kg,' to be answered readily. If the points are joined with straight lines rather than a curve, the diagram is called a *cumulative frequency polygon.*

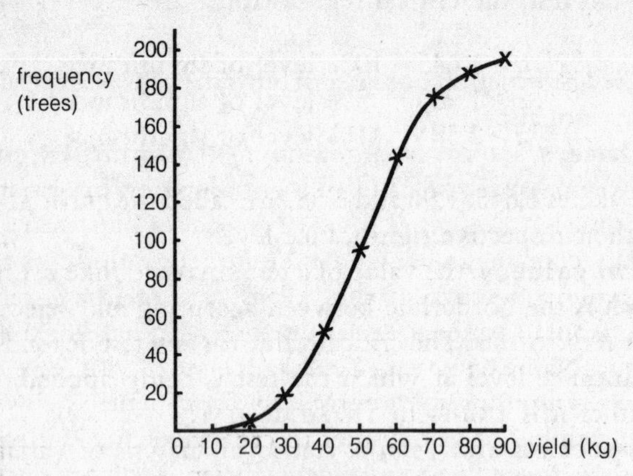

Fig. 22. **Cumulative frequency**

cyclic component, *n.* a non-seasonal component in a TIME SERIES which varies in a recognizable cycle. An example is the frequency of sunspots, which has an 11.2-year cyclic component (Fig. 23).

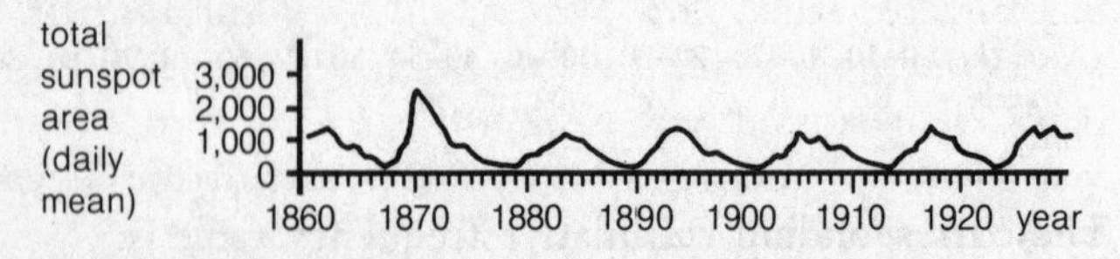

Fig. 23. **Cyclic component.** The 11.2-year cycle of solar activity (after W.J.S. Lockyer).

D

data, *n.* information obtained from a survey, an experiment or an investigation.

data base, *n.* a store of a large amount of information, especially in a form that can be handled by a computer. Information in a data base is stored as a number of records or files, each of which usually contains entries under a number of headings or fields. Example: a doctors' surgery keeps, as a data base, records of visits from its patients, under the following headings: Name of doctor; Name of patient; Sex; Age; Date of visit (day, month, year); Symptoms; Diagnosis; and Treatment.

The ability of a computer to cross-reference within such a data base makes it a particularly efficient form for storing and handling data. It could, for example, be made to search the data base for all patients aged between 35 and 45 diagnosed as suffering from varicose veins.

data capture, *n.* any process for converting information into a form that can be handled by a computer.

datum zero, *n.* a base level from which values of a variable are taken. For example, heights of tides are given with respect to a datum zero, called Low Water Ordinary Springs, (LWOS), and defined physically by a mark on Newlyn harbour wall; dates are measured with respect to a datum zero, the start of year 1 AD; longitude is measured with respect to the Greenwich meridian as datum zero.

In the ASSUMED MEAN method of working out averages, the assumed mean is a datum zero for the numbers involved.

death rate, *n.* the ratio of deaths in a specified area, group, etc., to the population of that area, usually expressed per thousand per year.

decile, *n.* one of nine actual or notional values of a variable dividing its distribution into 10 groups with equal frequencies.

decimal places, *n.* the figures to the right of the decimal point that are required to express the magnitude of a number to a specified degree of accuracy.
Example: 3.141,592,7 written to four decimal places is 3.1416; i.e., there are four digits after the decimal point, namely 1,4,1 and 6 (where the final digit represents a rounding-up or -down of any subsequent digits).

degrees of freedom, *n.* the number of free variables in a system. Symbol: ν. There are many situations where correct knowledge of the number of degrees of freedom is essential. These often involve the use of tables at the end of a calculation.

The idea of degrees of freedom can be seen from this obviously ridiculous argument. Lim and Tan are two Chinese boys. Their heights are 164cm and 156cm respectively, and in their end-of-term mathematics test they score 80% and 40%. Looking at these figures, someone concludes that 'Tall Chinese boys are better at mathematics than short ones'. He illustrates his theory on a graph (Fig. 24a). He then works out the CORRELATION COEFFICIENT, and finds it to be 1, providing (he claims) the final confirmation of his theory.

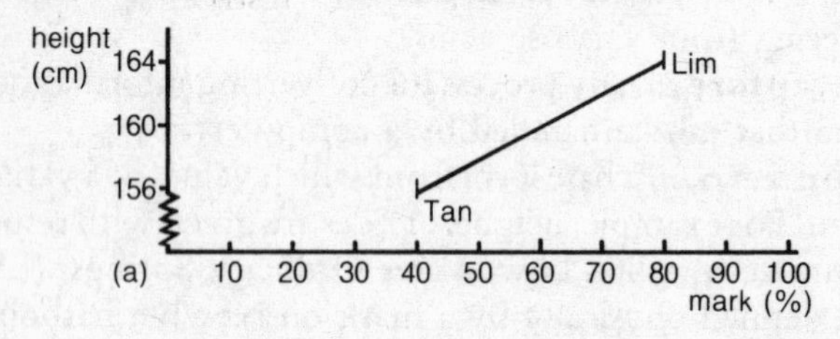

Fig. 24(a). **Degrees of freedom.**

The flaw in his argument is that if he takes only two points he will always get perfect correlation (positive or negative) because it takes two points to define his theory. Only when he takes a third, (and subsequent) pair of values is he in a position to carry out a test at all. To continue the example, suppose a third boy, Heng, was 162cm tall and scored 30% on the test.

When his point is added to the graph (Fig. 24b) it is not found to lie near the line defined by the other two. Although there are three points on the graph, two are used in defining the line (or theory); only one is actually being used in testing it. There are thus $3-2=1$ degrees of freedom in the system.

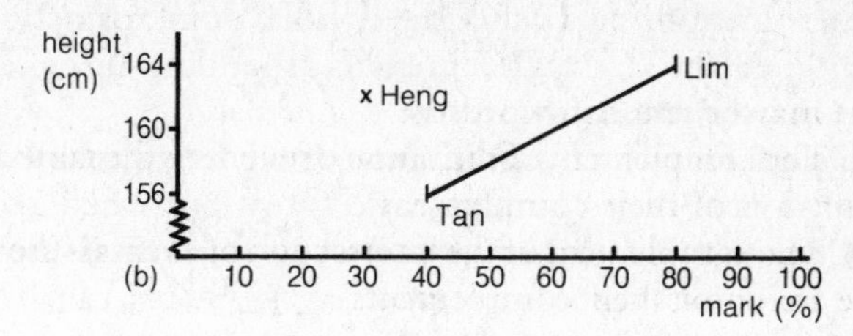

Fig. 24(b). **Degrees of freedom.**

In correlation generally, the number of degrees of freedom is similarly given by $\nu = n - 2$, where n is the number of points. It is not, however, usually possible to pick out two particular lost points. It is more appropriate to think of the set as a whole having lost two from its total of n independent variables. As a general rule, for n observations:

Degrees of freedom = n − restrictions

For the CHI-SQUARED TEST, with N groups, the number of degrees of freedom in different cases are given by:

Test for	ν
Normal distribution	$N-3$
Poisson distribution	$N-2$
Binomial distribution	$N-2$
Distribution in fixed proportion	$N-1$
u x v contingency table	$(u-1)(v-1)$

For the t-TEST, the number of degrees of freedom is $n-1$ (when comparing a sample mean against a parent mean) and $n_1 + n_2 - 2$ (when comparing two sample means).

demography, *n.* the science of population statistics.

de Morgan's laws, *n*. the laws which state that, for SETS or events $A_1, A_2, \ldots A_n$,

$$\text{(a)} \quad \left(\bigcup_{i=1}^{n} A_i\right)' = \bigcap_{i=1}^{n} A_i'$$

$$\text{(b)} \quad \left(\bigcap_{i=1}^{n} A\right)' = \bigcup_{i=1}^{n} A_i'$$

These may be stated in words as:

(a) 'The complement of the union of events is the same as the intersection of their complements,

(b) 'The complement of the intersection of events is the same as the union of their complements'.

These laws are illustrated by this simple example from sets.

$\mathscr{E} = \{a, b, c, d, e, f\}$, $A_1 = \{a, b, c\}$ and $A_2 = \{b, c, d\}$.

In this case, *de Morgan's first law* gives
$A_1 \cup A_2 = \{a, b, c, d\}$, and so $(A_1 \cup A_2)' = \{e, f\}$;
$A_1' = \{d, e, f\}$, $A_2' = \{a, e, f\}$, and so $A_1' \cap A_2' = \{e, f\}$;
so $(A_1 \cup A_2)' = A_1' \cap A_2'$.
de Morgan's second law gives
$A_1 \cap A_2 = \{b, c\}$, and so $(A_1 \cap A_2)' = \{a, d, e, f\}$.
$A_1' \cup A_2' = \{a, d, e, f\}$.
So $(A_1 \cap A_2)' = A_1' \cup A_2'$.

dependent variable, *n*. a variable which depends on another (the *independent*, or EXPLANATORY VARIABLE). When a graph is drawn, the dependent variable is usually drawn on a vertical (y) axis, the independent variable on the horizontal (x) axis (Fig. 25).

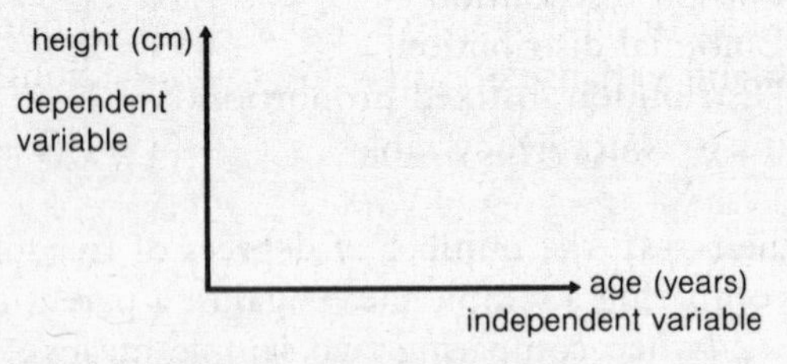

Fig. 25. **Dependent variable.**

If the graph were drawn of a girl's height, measured every birthday, against her age, her height would be the dependent variable, her age the independent. Her height depends on her age but her age does not depend on her height, just on the length of time she has been alive.

descriptive statistics, *n.* that branch of statistics which involves describing, displaying or arranging data; compare INFERENTIAL STATISTICS. PIE CHARTS, BAR CHARTS, PICTOGRAMS, etc., are all used in descriptive statistics. See also INTERPRETIVE STATISTICS.

deterministic model, *n.* a model in which all events are the inevitable consequences of antecedent causes.

deviation, *n.* the difference between a value of a variable and the mean of its distribution.

Example: the numbers 3, 6, 7, 10 and 4 have mean 6, and so their deviations are:

$$\begin{aligned} 3-6 &= -3 \\ 6-6 &= 0 \\ 7-6 &= 1 \\ 10-6 &= 4 \\ 4-6 &= -2 \end{aligned}$$

The sum of all the deviations in such a set should be zero.

Deviation is used in the calculation of VARIANCE, STANDARD DEVIATION, COVARIANCE, etc., but is not often used in its own right.

difference and sum of two or more random variables, *n.* those variables which are the difference and the sum of two or more random variables. When two random variables, X_1 and X_2, are added or subtracted to produce a third random variable, its mean and variance are given by these relationships.

Mean $(X_1+X_2) = \text{Mean } (X_1) + \text{Mean } (X_2)$ or $E(X_1+X_2) = E(X_1) + E(X_2)$
Variance $\text{Var}(X_1+X_2) = \text{Var}(X_1) + \text{Var}(X_2) + 2\text{cov}(X_1,X_2)$

Mean $(X_1-X_2) = \text{Mean } (X_1) - \text{Mean } (X_2)$ or $E(X_1-X_2) = E(X_1) - E(X_2)$
Variance $\text{Var}(X_1-X_2) = \text{Var}(X_1) + \text{Var}(X_2) - 2\text{cov}(X_1,X_2)$

where E is expectation, cov is covariance and Var is variance.

If X_1 and X_2 are independent variables then $\text{cov}(X_1,X_2)=0$. In that case

$$\text{Var}(X_1+X_2)=\text{Var}(X_1)+\text{Var}(X_2)$$
$$\text{Var}(X_1-X_2)=\text{Var}(X_1)+\text{Var}(X_2)$$

Example: a manufacturer makes knives with steel blades and wooden handles. The blades have mean mass 20g, standard deviation 1.5g; the handles have mean mass 15g, standard deviation 2g. What is the mean mass of a knife, and its standard deviation?

$$Mean=20+15=35\text{g}$$
$$SD=\sqrt{\text{Variance}}=\sqrt{1.5^2+2^2}=2.5\text{g}$$

Example: a joiner buys pieces of timber with mean length 300cm, SD 7.5cm. He cuts off pieces with mean length 250cm, SD 4cm. What are the mean and SD of the length of the offcuts?

$$Mean=300\text{cm}-250\text{cm}=50\text{cm}$$
$$SD=\sqrt{7.5^2+4^2}=8.5\text{cm}$$

These rules can be extended to more than two independent variables.

$$\text{E}(X_1 \pm X_2 \pm X_3 \pm \ldots) = \text{E}(X_1) \pm \text{E}(X_2) \pm \text{E}(X_3) \pm \ldots$$
$$\text{Var}(X_1 \pm X_2 \pm X_3 \pm \ldots) = \text{Var}(X_1)+\text{Var}(X_2)+\text{Var}(X_3)+\ldots$$

If the variables are not independent, the formula for variance must include terms $\pm\ 2\text{cov}(x_r,x_s)$ for all pairs r and s.

discrete, *adj.* consisting of distinct or separate parts.

discrete random variable, *n.* a variable which may take only certain discrete values; compare CONTINUOUS RANDOM VARIABLE. The number of people resident in a household is a discrete variable which may have value 1, 2, 3, etc., but not intermediate values, such as 1.5 or 2.446. On the other hand, the weight of the heaviest member of the household is a continuous variable, since it can take any reasonable value and is not restricted in this way.

disjoint, *adj.* (of SETS) non-overlapping. In the Venn diagram (Fig. 26) the sets A and B are disjoint, $A \cap B = \emptyset$.

The equivalent term for events (i.e. describing two events that cannot both occur) is EXCLUSIVE.

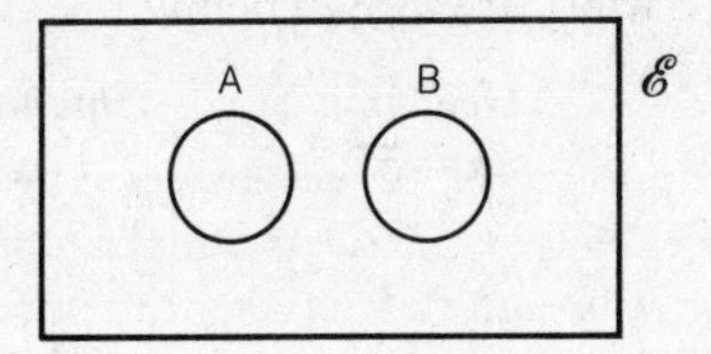

Fig. 26. **Disjoint.** A Venn diagram of disjoint sets A and B.

dispersion, *n.* the spread of a distribution. In the diagram (Fig. 27) distribution B has greater dispersion or spread than A although they both have the same mean.

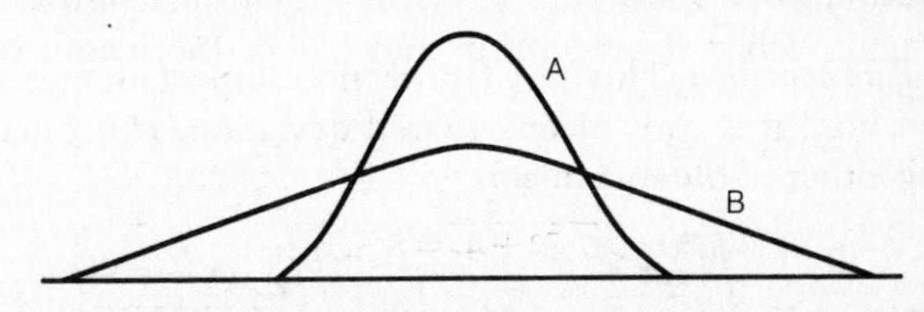

Fig. 27. **Dispersion.** Diagram of two distributions with the same mean, but different dispersions.

There are several measures of dispersion. Mean absolute deviation, standard deviation and semi-interquartile range are all measures of how far above or below the middle a typical member might be found. Standard deviation is the most important statistically and also, for small data sets, the most tedious to calculate by hand; it gives more weight to extreme values than do the other two.

For the set of numbers 65, 63, 58, 52, 47

Range = 18
Mean absolute deviation = 6
Standard deviation = 6.723
Interquartile range = 14.5
Semi-interquartile range = 7.25

Mean absolute deviation. The mean of the numbers, 65, 63, 58, 52, 47, is 57. The calculations of this, and of the mean absolution deviation, are shown below:

Number	*Deviation*	*Absolute deviation*
x	$x-57$	$\lvert x-57\rvert$
65	8	8
63	6	6
58	1	1
52	−5	5
47	−10	10
285		30

Mean $\bar{x}=57$
Mean absolute deviation = 6 (mean of the 5 values of the absolute deviation, $30 \div 5 = 6$).

Standard deviation. This is by far the most important measure of dispersion; it is root mean squared deviation. It is calculated using either of the formulae:

$$\text{(a)} \sqrt{\left\{\frac{\sum_i x_i^2}{n} - \bar{x}^2\right\}} \quad \text{(b)} \sqrt{\left\{\frac{\sum_i (x_i - \bar{x})^2}{n}\right\}}$$

The calculations for the two methods are illustrated below using the same figures as in the previous examples.

(a)	x	x^2
	65	4,225
	63	3,969
	58	3,364
	52	2,704
	47	2,209
	285	16,471

$n=5$

Mean $\bar{x}=57$

$$\textit{Standard deviation} = \sqrt{\left\{\frac{16{,}471}{5} - 57^2\right\}} = 6.723$$

(b)

x	$(x-\bar{x})$	$(x-\bar{x})^2$
65	8	64
63	6	36
58	1	1
52	-5	25
47	-10	100
285		226

$n=5$

Mean $\bar{x}=57$

Standard deviation $= \sqrt{\dfrac{226}{5}} = 6.723$

These two formulae can also be written:

$$\sqrt{\left\{\sum_i \frac{f_i x_i^2}{n} - \bar{x}^2\right\}} \qquad \sqrt{\sum_i \frac{f_i (x_i-\bar{x})^2}{n}}$$

for use with grouped data.

Range. This is the difference between the largest and the smallest of a set of variables. Thus the range of the values 65, 63, 58, 52, 47 is $65-47=18$.

Interquartile range. This is the difference between the upper and lower quartiles. *Semi-interquartile range* is half of the interquartile range. These measures are best applied to large rather than small sets of figures. It is, however, possible to use the same set as in the other examples.

x	*Rank*	
65	1	←Upper quartile = 64 (between ranks 1 and 2)
63	2	
58	3	
52	4	←Lower quartile = 49.5 (between ranks 4 and 5)
47	5	

Interquartile range $=64-49.5=14.5$

Semi-interquartile range $= \dfrac{14.5}{2} = 7.25$

The standard deviation of a normal distribution may also be estimated from the range of a sample using the table on page 257. With the figures used in the examples (65, 63, 58, 52, 47) the range is 18, and n=5. The standard deviation is then estimated as:

$$a_5 \times \text{range} = .4299 \times 18 = 7.74$$

where $a_5 = .4299$ is obtained from the table. This can be compared with 7.52, the estimate of parent standard deviation found by using the formula

$$\sqrt{\left[\frac{\Sigma_i (x_i - \bar{x})^2}{n-1}\right]}$$

on the five sample figures.

distribution, *n.* the set of frequencies (*observed distribution*) or probabilities (*theoretical distribution*) assigned to a set of events. Example: *frequency distribution* (in which frequencies are given). The lengths of the reigns of English kings and queens, 827–1952 AD.

Length (years)	<5	5≤<10	10≤<15	15≤<20	20≤<25
Frequency	11	11	8	8	7

Length (years)	25≤<30	30≤<35	35≤<40	40≤<45
Frequency	4	1	6	1

Length (years)	45≤<50	50≤<55	55≤<60	60≤<65
Frequency	0	1	2	1

Example: PROBABILITY DISTRIBUTION (in which probabilities are given). The probability distribution of different totals when two dice are thrown. The top line (below) represents the possible combined scores of two dice. The bottom line shows the probability of each combined score occurring.

Total score	2	3	4	5	6	7	8	9	10	11	12
Probability	$\frac{1}{36}$	$\frac{2}{36}$	$\frac{3}{36}$	$\frac{4}{36}$	$\frac{5}{36}$	$\frac{6}{36}$	$\frac{5}{36}$	$\frac{4}{36}$	$\frac{3}{36}$	$\frac{2}{36}$	$\frac{1}{36}$

Figs. 28, 29, 30, 31, 32a and b show the graphs of some common distributions.

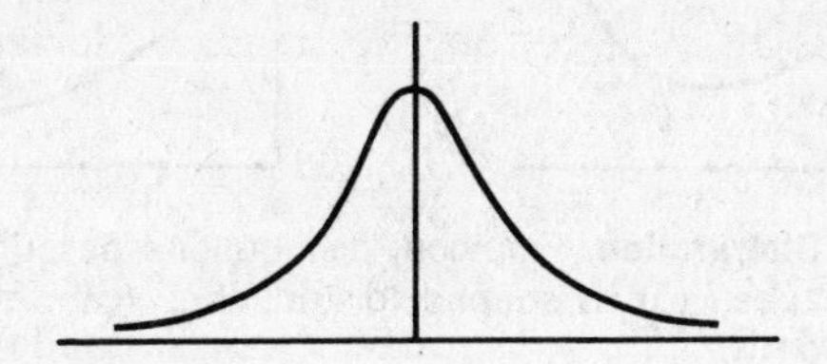

Fig. 28. **Distribution.** Normal distribution; the vertical line represents the mean value.

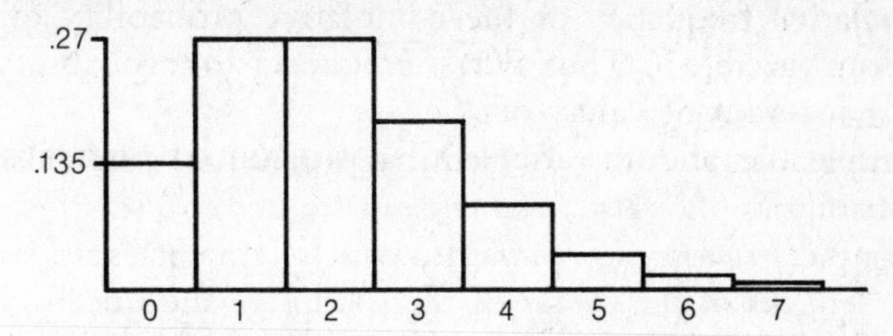

Fig. 29. **Distribution.** Poisson distribution; in this example, the mean μ is 2, so the distribution is denoted P(2).

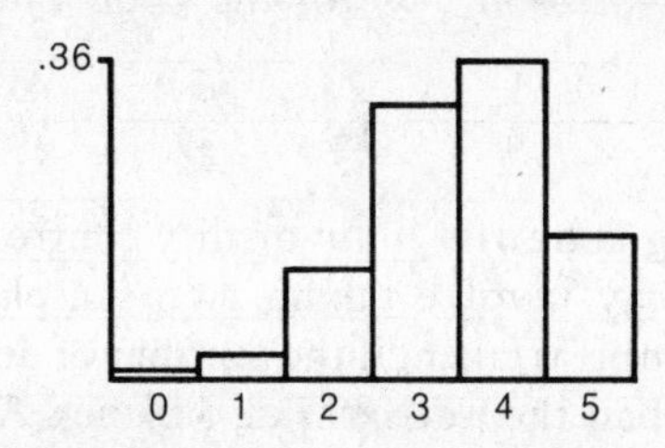

Fig. 30. **Distribution.** Binomial distribution; in this example, the number of trials n is 5, the probability of success p is .7. Thus the distribution is denoted B(5, .7).

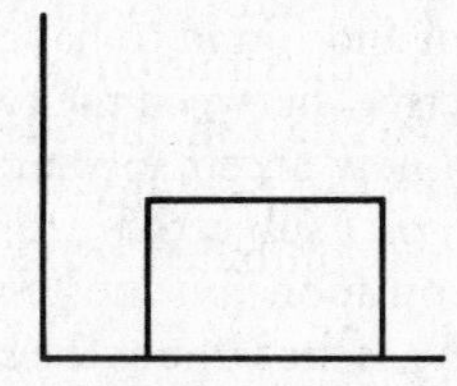

Fig. 31. **Distribution.** Uniform distribution.

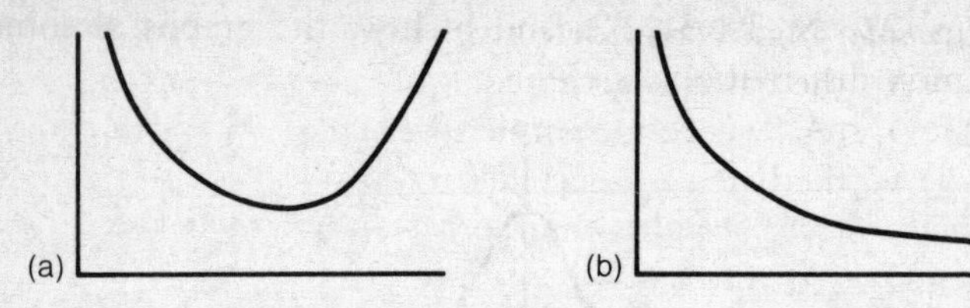

Fig. 32. **Distribution.** Empirical distributions described by their shapes. (a) U-shaped distribution. (b) J-shaped distribution.

distribution function, *n.* the function $F(x)$ which gives the cumulative frequency or the cumulative probability of the random variable X. Thus, $F(x)$ = Frequency (or probability) of $X \leqslant x$ for a set of values of X.

Example: the random variable X has PROBABILITY DISTRIBUTION $p(x)$:

x	0	1	2	3	4	5
$p(x)$	$\frac{1}{32}$	$\frac{5}{32}$	$\frac{10}{32}$	$\frac{10}{32}$	$\frac{5}{32}$	$\frac{1}{32}$

(i.e. the binomial distribution $p = \frac{1}{2}$, $n = 5$).

The distribution function $F(x)$ for the same values, is thus:

x	0	1	2	3	4	5
$F(x)$	$\frac{1}{32}$	$\frac{6}{32}$	$\frac{16}{32}$	$\frac{26}{32}$	$\frac{31}{32}$	$\frac{32}{32}$

double sampling scheme, *n.* (in quality control) a sampling scheme which may involve taking two samples. In such a scheme, a first sample is taken. If the number of defective items D_1 is not more than the ACCEPTANCE NUMBER A_1, the whole batch is accepted. If it is greater than another, larger number A_2, the batch is rejected (or subjected to 100% inspection). If the number of defectives lies between these two numbers, a second sample is taken and found to have D_2 defectives. If the total number of defectives between the two samples, $D_1 + D_2$, is then greater than a new acceptance number A_3, the whole batch is rejected or subjected to 100% inspection ($A_1 < A_2 \leqslant A_3$). It is common (but not essential) that the new acceptance number, A_3, is the same as the larger number for the first stage of sampling, A_2.

First sample: Number of defectives is D_1.
If $D_1 \leqslant A_1$, the batch is accepted.
If $A_1 < D_1 \leqslant A_2$, a second sample is taken.
If $D_1 > A_2$, the batch is rejected.

Second sample: Number of defectives is D_2.
If $D_1 + D_2 \leqslant A_3$, the batch is accepted.
If $D_1 + D_2 > A_3$, the batch is rejected.

E

econometrics, *n.* the application of mathematical and statistical techniques to economic theories.

effect variable, see EXPLANATORY VARIABLE.

efficient estimator, *n.* among two or more ESTIMATORS, that one with the smaller variance.

Example: in an experiment, three independent readings x_1, x_2 and x_3 are taken of a random variable with mean μ and variance V. Which is the most efficient estimator for μ, (a), (b), or (c)?

(a) $\dfrac{x_1+x_2}{2}$ (b) $\dfrac{2x_1+x_2+x_3}{4}$ (c) $\dfrac{x_1+x_2+x_3}{3}$

All three are unbiased estimators since

$$\text{(a)}\quad \mathrm{E}\left(\frac{x_1+x_2}{2}\right)=\tfrac{1}{2}\mathrm{E}(x_1)+\tfrac{1}{2}\mathrm{E}(x_2)=\tfrac{1}{2}\mu+\tfrac{1}{2}\mu=\mu$$

$$\text{(b)}\quad \mathrm{E}\left(\frac{2x_1+x_2+x_3}{4}\right)=\tfrac{2}{4}\mathrm{E}(x_1)+\tfrac{1}{4}\mathrm{E}(x_2)+\tfrac{1}{4}\mathrm{E}(x_3)$$
$$=\tfrac{1}{2}\mu+\tfrac{1}{4}\mu+\tfrac{1}{4}\mu=\mu$$

$$\text{(c)}\quad \mathrm{E}\left(\frac{x_1+x_2+x_3}{3}\right)=\tfrac{1}{3}\mathrm{E}(x_1)+\tfrac{1}{3}\mathrm{E}(x_2)+\tfrac{1}{3}\mathrm{E}(x_3)$$
$$=\tfrac{1}{3}\mu+\tfrac{1}{3}\mu+\tfrac{1}{3}\mu=\mu$$

Their variances are calculated as follows:

$$\text{(a)}\quad \mathrm{Var}\left(\frac{x_1+x_2}{2}\right)=\mathrm{Var}\left(\frac{x_1}{2}\right)+\mathrm{Var}\left(\frac{x_2}{2}\right)$$
$$=\tfrac{1}{4}\mathrm{Var}(x_1)+\tfrac{1}{4}\mathrm{Var}(x_2)=\tfrac{1}{2}\mathrm{V}$$

(b) $\text{Var}\left(\frac{2x_1 + x_2 + x_3}{4}\right)$

$$= \text{Var}\left(\frac{x_1}{2}\right) + \text{Var}\left(\frac{x_2}{4}\right) + \text{Var}\left(\frac{x_3}{4}\right)$$

$$= \tfrac{1}{4}\text{Var}(x_1) + \tfrac{1}{16}\text{Var}(x_2) + \tfrac{1}{16}\text{Var}(x_3)$$

$$= \tfrac{3}{8}\text{V}$$

(c) $\text{Var}\left(\frac{x_1 + x_2 + x_3}{3}\right)$

$$= \text{Var}\left(\frac{x_1}{3}\right) + \text{Var}\left(\frac{x_2}{3}\right) + \text{Var}\left(\frac{x_3}{3}\right)$$

$$= \tfrac{1}{9}\text{Var}(x_1) + \tfrac{1}{9}\text{Var}(x_2) + \tfrac{1}{9}\text{Var}(x_3)$$

$$= \tfrac{1}{3}\text{V}$$

It is seen that, as $\frac{1}{3}$V is the smallest variance, (c) is the most efficient estimator of the three.

Eigen values, Eigen vectors, see TRANSITION MATRIX.

empirical, *adj*. derived from or relating to experiment and observation, rather than theory. An example of an empirical law is the Rank Size Rule, used in geography. This states that if P_1 is the population of a country's largest city (or town), then P_1/n is that of its nth city. There is no theoretical basis for this rule; it has just been found to work in many cases, as in the case of the USSR (1979 figures).

Rank	*City*	*Actual size (× 1000)*	*Predicted size (× 1000) (Rank Size Rule)*
1	Moscow	8,011	8,011
2	Leningrad	4,588	4,006
3	Kiev	2,144	2,670
4	Tashkent	1,779	2,003
5	Baku	1,550	1,602
6	Kharkov	1,444	1,335

Source: *Philip's Certificate Atlas*, 1978.

An example of a law which is not empirical is that which states that when two dice are thrown repeatedly the possible scores 2, 3, 4, 5, 6, 7, 8, 9, 10, 11, 12, will occur with frequencies in the ratio.

$$1 : 2 : 3 : 4 : 5 : 6 : 5 : 4 : 3 : 2 : 1$$

This is based upon the fact that these are the numbers of ways that the different scores can be attained, as shown in the table below.

First die						
6	7	8	9	10	11	12
5	6	7	8	9	10	11
4	5	6	7	8	9	10
3	4	5	6	7	8	9
2	3	4	5	6	7	8
1	2	3	4	5	6	7
	1	*2*	*3*	*4*	*5*	*6*
	Second die					

enumeration data, *n.* data consisting of numbers of individuals in various categories.
Example: a flower has three varieties, red, white and pink. 400 flowers are collected and found to be:

Red	89
Pink	201
White	110

There are thus three categories, red, pink and white; the data consist of the number of flowers in each.

equiprobable, *adj.* (of values) of equal probability. The possible scores when a die is thrown, 1, 2, 3, 4, 5, and 6, are equiprobable, each has a probability of occurring of 1/6. The scores 0 to 36 on a fair French roulette wheel, are also equiprobable, at 1/37 each.

error, *n.* **1.** a mistake or inaccuracy. **2.** the magnitude of an inaccuracy. See also RANDOM ERROR.

error curve, *n.* the NORMAL DISTRIBUTION curve.

estimate, 1. *n.* an approximate calculation. **2.** *n.* an approximate idea of something (e.g., size, cost). **3.** *vb.* to make an approximate calculation, or to form an approximate idea of something (e.g., size, cost).

estimator, *n.* a random variable, formula or procedure for estimating the value of a parent population PARAMETER from sample data. Its value depends on the particular sample involved. An estimator should ideally be consistent, unbiased and efficient. Estimated values of population parameters are distinguished from their true values by the convention of using the symbol ^. Thus σ = true parent population standard deviation while $\hat{\sigma}$ = estimated (from a sample) parent population standard deviation. See also EFFICIENT ESTIMATOR, LINEAR ESTIMATOR.

Euler diagram, see VENN DIAGRAM.

event, *n.* something which may or may not occur. Examples:
a coin showing heads when tossed;
a particular seed germinating;
a motorist being caught speeding on a particular journey.

exclusive, *adj.* (of two events) such that both events cannot occur. Thus events A and B are exclusive if B cannot occur when A does, and vice versa. It is, however, possible that neither A nor B occurs.
Example:

A: getting a 6 when throwing a die
B: getting a 4 on the same throw of the die

When three or more events are such that no two of them can occur at the same time, they are said to be *mutually exclusive.* Compare DISJOINT.

exhaustive, *adj.* **1.** (of two events) such that at least one must occur. Thus events P and Q are exhaustive if either P alone, or Q alone, or both P and Q together, *must* occur; it is not possible for neither to occur. Exhaustive events cover all possibilities.
Example:

P: An athlete does a track event
Q: An athlete does a field event

Since all athletics events are track, or field, or both (e.g. the decathlon), P and Q are exhaustive events. **2.** (of a greater number of events) such that they cover all possibilities between them, they are comprehensive in their scope.

expectation, *n.* the expected or MEAN value of a random variable, or of a function of the variable. Symbol: E( ). For a DISCRETE VARIABLE, X, taking values $x_1, x_2, \ldots, x_n$ with probabilities $p(x_1), p(x_2), \ldots, p(x_n)$, the expectation of X, $E(X)$ or $\bar{X}$, is given by:

$$\sum_{i=1}^{i=n} x_i p(x_i)$$

and for a CONTINUOUS VARIABLE

$$\int x f(x)\, dx$$

where $f(x)$ is the PROBABILITY DENSITY FUNCTION of X.
Example: when a die is thrown, each of the possible values 1, 2, 3, 4, 5 and 6 has probability of 1/6 of showing. So the expectation of the value shown is

$$1 \times \tfrac{1}{6} + 2 \times \tfrac{1}{6} + 3 \times \tfrac{1}{6} + 4 \times \tfrac{1}{6} + 5 \times \tfrac{1}{6} + 6 \times \tfrac{1}{6} = 3.5$$

Expectation obeys the following laws:

$$\begin{aligned} E(\lambda X) &= \lambda E(X) \\ E(X_1 \pm X_2) &= E(X_1) \pm E(X_2) \\ E(X^2) &= [E(X)]^2 + \mathrm{Var}(X) \\ E(X_1 X_2) &= E(X_1)E(X_2) + \mathrm{cov}(X_1, X_2) \end{aligned}$$

where Var is variance; cov is covariance.

If $g(X)$ is a function of the random variable X with probability density function $f(x)$, then

$$E(g(X)) = \int g(x) f(x)\, dx$$

so that

$$\begin{aligned} E(X^m) &= \int x^m f(x)\, dx \\ E(\sin x) &= \int \sin x\, f(x)\, dx \end{aligned}$$

experiment, *n.* any process which results in the collection of data. (This meaning for the word is somewhat different from that usually given to it in other scientific disciplines.) Examples of experiments in a statistical sense: carrying out a survey; measuring the daily rainfall; recording the longevity of animals of a particular type.

experimental design, *n.* the way in which a statistical EXPERIMENT is planned. The example which follows illustrates the use of a number of different designs, namely: *completely randomized design, randomized blocks design, Latin square,* and *Graeco-Latin square.*
Example: a market gardener wants to test three types of peas, A, B and C, on his land. He has a square plot which he divides into nine equal squares, three to be planted with each type of pea. The problem which he then faces is which square to plant with which type. One method is complete randomization which might, for example, result in the pattern in Fig. 33.

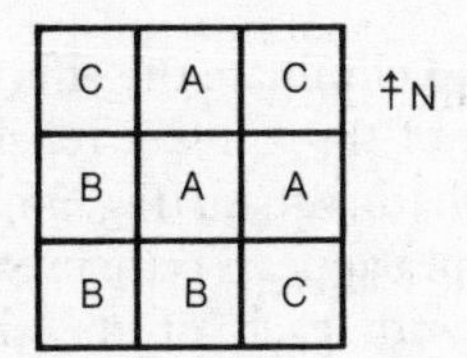

Fig. 33. **Experimental design.** Completely randomized design for a square with nine equally-sized plots, in a test for three types of pea, A, B, C.

block 1	A	B	C	↑N
block 2	A	C	B	
block 3	C	B	A	

Fig. 34. **Experimental design.** A square plot divided for a randomized block design for three types of pea, A, B, C.

This would be all right if all the plots were equally desirable. If, however, there were a prevailing north wind so that the northernmost plots were more exposed, he might decide to use randomized blocks (Fig. 34), where each of the types A, B and C is planted once in each west-east block.

If the gardener also felt that the soil to the east was rather better than that to the west, he would use a Latin square (Fig. 35), where each type of pea were planted once in each row, and once in each column.

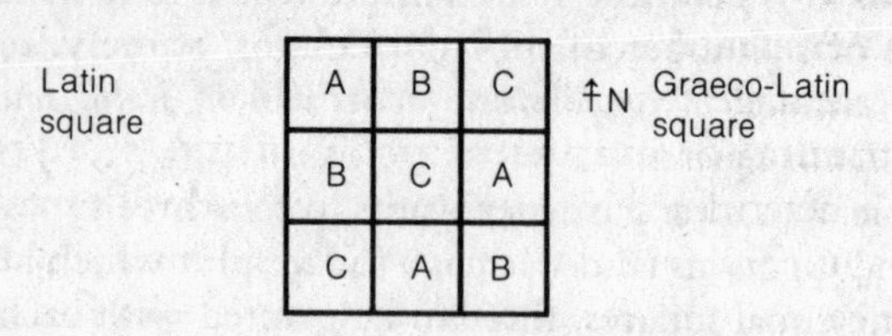

Fig. 35. **Experimental design.** A square plot divided into a Latin square, to test three types of pea, A, B, C, where each type appears once in each row, and once in each column.

If he also wished to eliminate the effects of a third variable, for example the use of three insect repellents, α, β and γ, he would use a Graeco-Latin square (Fig. 36). In this experimental design, each type of pea appears once in each row, once in each column and once with each of the three types of insect repellent.

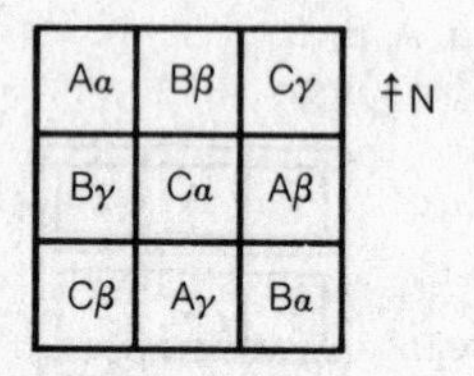

Fig. 36. **Experimental design.** A square plot divided into a Graeco-Latin square to eliminate the effects of a third variable, α, β, γ.

A Graeco-Latin square is really two Latin squares, one superimposed on the other. There are Graeco-Latin squares of sides 3, 4, 5, 7, 8, 9 and 10, but not 6.

experimental error, *n.* the difference between the value of a quantity found in an experiment and that which should have been found had the experiment been conducted perfectly. Experimental error is due to experimental technique, and does not include mistakes in mathematics or EXPERIMENTAL DESIGN. Example: a physics student conducting an experiment to find the gravitational acceleration g, gets an answer of $9.99ms^{-2}$ instead of $9.81ms^{-2}$.

The experimental error is $0.18ms^{-2}$.

The percentage experimental error is 1.83%.

experimental or **empirical probability,** *n.* a probability which is determined by experiment or observation rather than from some underlying theory. The figures used by life insurance companies in calculating the risks for particular individuals come into this category. On the other hand, a skilled bridge player decides his strategy on the basis of theoretical probabilities.

explanatory variable *n.* **1.** or **cause variable,** a variable upon which another depends. If one variable y is caused by another x, then x is called the explanatory variable, and y the *effect variable*. If, for example, cigarette smoking is regarded as causing lung cancer, then the number of cigarettes smoked per day could be the cause variable, the probability of death by lung cancer the effect variable. **2.** or **independent variable** (see DEPENDENT VARIABLE), when the idea of cause may or may not be involved.

exploratory data analysis, *n.* the process of looking at raw data to decide on its important features. It may involve such procedures as:

(a) ROUNDING the figures, or cutting out unnecessary numbers of significant figures;

(b) Grouping the data in a convenient form, such as, for example, a STEM-AND-LEAF PLOT or a HISTOGRAM; if the data came from several batches, BOX-AND-WHISKER PLOTS may be drawn to compare the batches for any central tendency and spread;

(c) Identifying OUTLIERS;

(d) Finding the MEDIAN and QUARTILE values.

exploratory survey, see PILOT SURVEY.

exponential, *adj*. (of a value or function) raised to the power of. Thus the function $F(x) = a^x$ is an exponential function; x is called the *exponent*.

The term exponential is most commonly used when the exponent is the power of e, the base of natural logarithms. In this case the exponential of x is e^x, which is sometimes written as $\exp(x)$.

exponential distribution, *n*. a continuous distribution with probability density function given by:

$$f(x) = \lambda e^{-\lambda x}$$

(see Fig. 37)

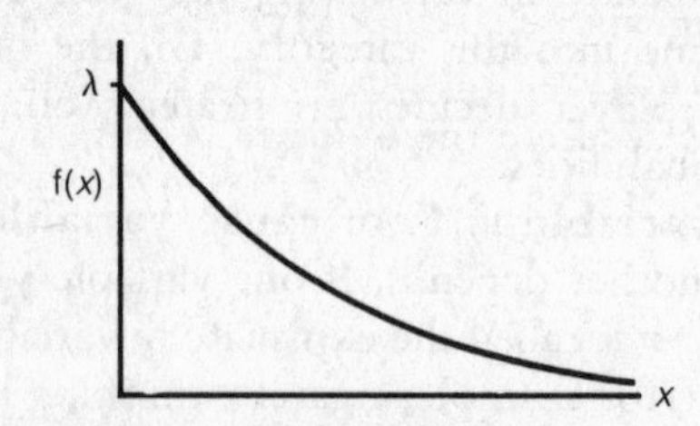

Fig. 37. **Exponential distribution.** For this distribution:

$$Mean = \frac{1}{\lambda} \qquad Variance = \frac{1}{\lambda^2}$$

The exponential distribution is sometimes used in connection with the length of life of certain materials.

extrapolation, *n*. estimation of a value of a variable beyond known values. Compare INTERPOLATION.

Example: Sheuli was 1.20m tall on 1 January 1975, and 1.40m on 1 January 1980. Estimate her height on 1 January 1985 by extrapolation.

From the graph, Fig. 38, it could be estimated by extrapolation that by 1 January 1985, she would have grown another 0.20m to be 1.60m tall. This assumes that she continued to grow at the same rate. Eventually this must become a false assumption, otherwise by 1 January 2008, she would be a record-breaking giantess of 2.52m.

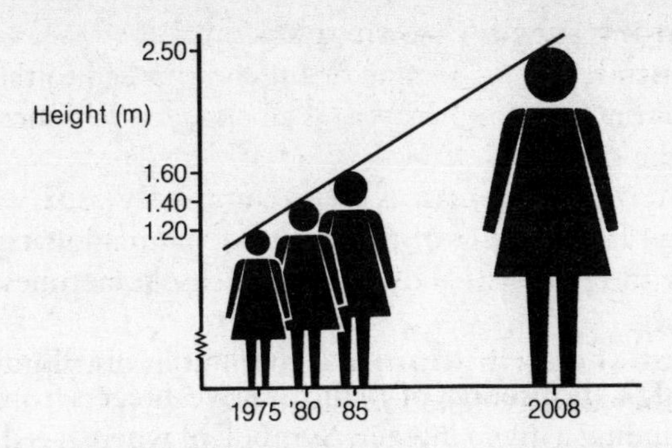

Fig. 38. **Extrapolation.** Graph of Sheuli's height in 1975 and 1980, with her height extrapolated for 1985, and 2008.

Extrapolation may be reasonably reliable near the known points of information, but the further away from them one works, the less accurate the estimate becomes.

F

factorial, *n.* the product of all the positive integers from 1 up to and including a given integer. Symbol: n! where n is the given integer. Thus 5 factorial, written 5!, means

$$5 \times 4 \times 3 \times 2 \times 1 \ (= 120)$$

Factorials are used in working out probabilities in the POISSON DISTRIBUTION and BINOMIAL DISTRIBUTION, in PERMUTATIONS and COMBINATIONS, etc.

Factorials very quickly become large numbers. By convention, 0! is given the value 1.

$$\begin{aligned} 1! &= 1 \\ 2! &= 2 \\ 3! &= 6 \\ 4! &= 24 \\ 5! &= 120 \\ 6! &= 720 \\ 7! &= 5{,}040 \\ 8! &= 40{,}320 \\ 9! &= 362{,}880 \\ 10! &= 3{,}628{,}800 \end{aligned}$$

For large values of n, n! is given approximately by

$$n! \simeq \sqrt{2\pi n}\ n^n e^{-n}, \qquad \text{or } n! \simeq \sqrt{2\pi}\ n^{(n+\frac{1}{2})} e^{-n}$$

which can be rewritten as *Stirling's approximation to n!*

$$\log_{10} n! = .39909 + (n + \tfrac{1}{2}) \log_{10} n - .4342945\ n$$

factorial experiment, *n.* an experiment designed to provide information about individual factors and the interaction between them. It leads up to calculations for ANALYSIS OF VARIANCE.

fertility rate, *n.* the ratio of live births in a specified area, group, etc., to the female population between the ages of 15 and 44, usually expressed per thousand per year.

finite population correction, *n.* a correction factor applied to the expression for the variance of the means of samples drawn from an infinite parent population, to allow for it being of finite size. When samples of size n are taken from an infinite parent population with standard deviation σ (or from a finite one with replacement) the variance and standard deviation of the sample means are given by:

$$\text{Variance} = \frac{\sigma^2}{n} \qquad \text{SD} = \frac{\sigma}{\sqrt{n}}$$

If, however, the samples are taken from a population of size N, without replacement, the variance and standard deviation of the sample means are given by:

$$\text{Variance} = \frac{\sigma^2}{n}\left(\frac{N-n}{N-1}\right) \qquad \text{SD} = \frac{\sigma}{\sqrt{n}}\sqrt{\frac{N-n}{N-1}}$$

The difference between these two expressions for the variance is the factor

$$\frac{N-n}{N-1}$$

which is the finite population correction.

If N is large in comparison with n, the finite population correction may be ignored.

Fisher's Ideal Index, see INDEX NUMBER.

Fisher's z-transformation, *n.* a transformation used in certain calculations involving CORRELATION COEFFICIENTS. Symbol: z. It is given by

$$z = \tfrac{1}{2}\ln\left(\frac{1+r}{1-r}\right) = \text{inv tanh } r$$

Fisher's z-transformation is used:

(a) In testing the hypothesis that the correlation coefficient of the parent population of a BIVARIATE DISTRIBUTION has a particular value:

(b) In setting up CONFIDENCE LIMITS for an estimated value of the parent population's correlation coefficient.

When a correlation coefficient r is calculated, the answer obtained is based on a sample. If, for example, 20 Siamese cats were used in a dietary experiment, the correlation coefficient between their masses (in kg) and daily food intake (in calories) could be calculated, giving a value of, say, 0.6. The 20 cats are a sample from a very large parent population, all Siamese cats. Thus the value of 0.6 is only an estimate of the parent population's correlation coefficient, based on a sample of size 20.

When significance level tables are used for interpreting correlation coefficients, the NULL HYPOTHESIS being tested is: There is no correlation. The sample is drawn from a parent population with correlation coefficient 0.

The alternative hypothesis is that the parent correlation coefficient is not zero. Thus a two-tail test is involved (see ONE- AND TWO-TAIL TESTS). In order to test this hypothesis, it should be necessary to know the distribution of correlation coefficients for samples of size 20. However, Fisher's transformation overcomes this problem by converting this distribution into a normal one. If r is a particular sample correlation coefficient, the corresponding value of z is:

$$z = \tfrac{1}{2}\ln\left(\frac{1+r}{1-r}\right)$$

The distribution of z is normal, with:

$$\textit{Mean} = z_0$$

$$\textit{Standard deviation} = \frac{1}{\sqrt{n-3}}$$

Thus, if ρ_0 is the true parent correlation coefficient, z_0 (the mean of the transformed distribution) is found from the equation above as:

$$z_0 = \tfrac{1}{2} \ln\left(\frac{1+\rho_0}{1-\rho_0}\right)$$

In the case of the 20 Siamese cats, the null hypothesis was that $\rho_0 = 0$, giving $z_0 = 0$. The z value (using $r = .6$) is

$$\tfrac{1}{2} \ln\left(\frac{1+.6}{1-.6}\right) = .693$$

This calculation of the z value can be avoided by using the tables on page 257, the Fisher z-transformation.

The standard deviation σ for $n = 20$ is:

$$\frac{1}{\sqrt{20-3}} = .243$$

Applying the normal distribution to these data (Fig. 39), the shaded areas are each $2(1' - \Phi(2.85)) = .0044$. So, if the null hypothesis, that $\rho_0 = 0$, is true, the probability of obtaining a value of r at least as far away from that value as the one observed (i.e. $r \geqslant .6$, or $r \leqslant -.6$) is .0044, the SIGNIFICANCE LEVEL of the result. There is thus good reason to reject the null hypothesis of no correlation.

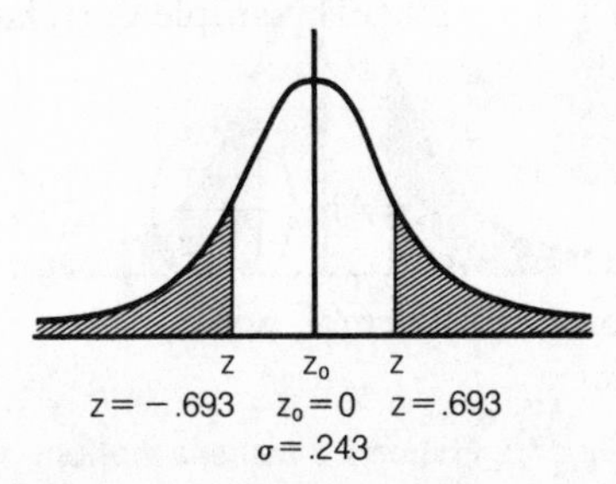

Fig. 39. **Fisher's z-transformation.** The transformed distribution. Standardizing the value of z, i.e. .693, gives:

$$\frac{.693 - 0}{.243} = 2.85$$

This answer, that the correlation coefficient is significant at better than the .5% significance level, could have been obtained from the table on page 254 without the use of Fisher's transformation.

However, Fisher's transformation does allow other null hypotheses to be tested. If, in the case of the Siamese cats, it is believed that the parent correlation coefficient is .5, this can be used as the null hypothesis.

In this example,

$$\rho_0 = .5 \Rightarrow z_0 = .549$$
$$r = .6 \Rightarrow z = .693$$

$$\sigma = \frac{1}{\sqrt{20-3}} = .243$$

Thus, the probability of a result as far or further from the mean (Fig. 40) is:

$$2\left(1 - \Phi\left[\frac{.693 - .549}{.243}\right]\right) = .556$$

This is a high probability, and so the result is likely. There is no reason to reject the null hypothesis that $\rho_0 = .5$.

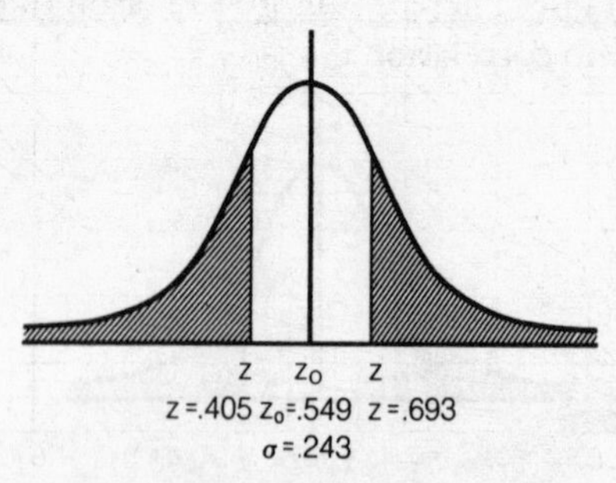

Fig. 40. **Fisher's z-transformation.**

It is, however, normally unlikely that the parent correlation coefficient is known; (that would mean the research had already been carried out). A more likely situation is that the sample is used to make an estimate of the parent correlation coefficient. In that case it would also be helpful to have CONFIDENCE LIMITS for the value obtained. In the example of the Siamese cats, the estimated value of z_0 would be .693, the transformed value of r=.6. The 95% confidence limits (see tables, p. 247) would then be worked out using the normal distribution:

$$z = z_0 \pm k\sigma = .693 \pm 1.96 \times .243$$
$$z = .217 \text{ and } 1.169$$

When these values of z are transformed back to r, using the *inverse Fisher transformation*, r = tanh z, they give r = .214 and .824.

These are the 95% confidence limits for ρ_0. Confidence limits for ρ may also be found using charts such as that in Fig. 41. In the example of the Siamese cats, r is estimated to be .6, and the sample size n is 20. On the chart, the line, r = .6 intersects both

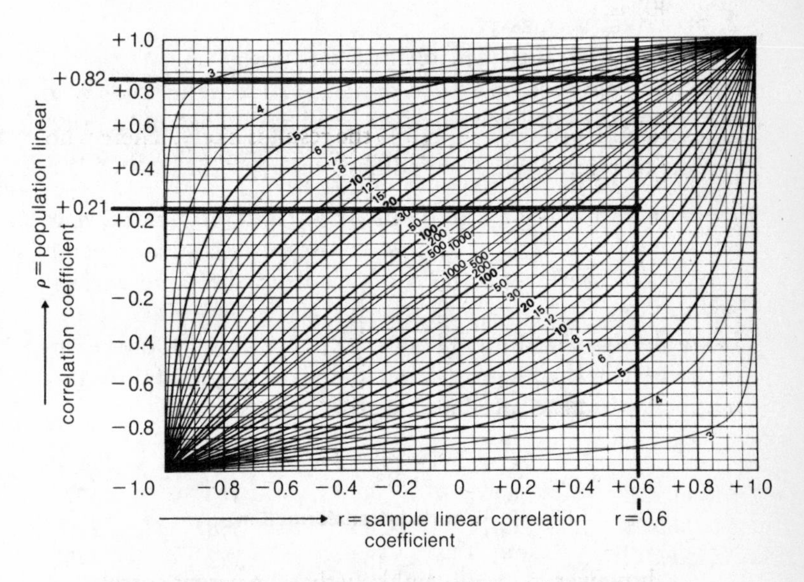

Fig. 41. **Fisher's z-transformation.** 95% confidence interval chart for correlation coefficients. Source: *Elementary Statistics Tables*, H.R. Neave, George Allen and Unwin.

curves for n = 20; the values of ρ corresponding to these two intersections are seen to be .21 and .82, the confidence limits for the population correlation coefficient.

frame, *n.* a representation of the items available to be chosen for a sample. This could be a telephone directory, a map, an electoral register, a list of the cattle in a herd, etc.

frequency, *n.* **1.** or **absolute frequency,** the number of individuals in a class. Symbol: f.

Example: in 20 shots at a target, a marksman makes the following scores:

4	3	2	4	5	2	4	2	5	1
0	5	1	5	2	5	5	0	1	4

The frequencies of the different scores are thus:

Score	0	1	2	3	4	5
Frequency	2	3	4	1	4	6

2. RELATIVE FREQUENCY.

frequency distribution, see DISTRIBUTION.

frequency surface, *n.* a three-dimensional representation of a SCATTER DIAGRAM. The area of the graph is divided into a grid, each part of which is raised to a height proportional to the frequency of the points lying within it.

Friedman's two-way analysis of variance by rank, *n.* a non-parametric test of the NULL HYPOTHESIS that several MATCHED SAMPLES have been drawn from the same PARENT POPULATION.

Example: in a trial, four operators (Op. 1, 2, 3 and 4) used each of three screw-making machines for a fixed length of time; the numbers of screws produced were:

	Op. 1	*Op. 2*	*Op. 3*	*Op. 4*
Machine A	361	385	340	377
Machine B	264	283	358	259
Machine C	420	434	392	445

Is there any evidence to suggest differences in the performances of the operators at the 5% SIGNIFICANCE LEVEL?

Null hypothesis: There is no difference in the operators' performance.
The figures in each row of the table are ranked (1 to 4) for the four operators, and the sums of the ranks for each calculated (R_i).

	Op. 1	*Op. 2*	*Op. 3*	*Op. 4*
Machine A	3	1	4	2
Machine B	3	2	1	4
Machine C	3	2	4	1
Total (R_i)	9	5	9	7

The value of the statistic M is then worked out using the formula:

$$M = \frac{12}{NK(K+1)} \sum_{i=1}^{K} R_i^2 - 3N(K+1)$$

where N is the number of machines (3), K is the number of operators (4), and R_i is the total of the ranks for operator i. Thus

$$M = \frac{12}{3 \times 4 \times (4+1)}(81+25+81+49) - 3 \times 3 \times (4+1) = 2.2$$

This statistic is looked up in the χ^2 tables for $\nu = K - 1$ $(=3)$ degrees of freedom (p. 249). At the 5% level, the critical value is 7.81. Since $2.2 < 7.81$, there is no reason to reject the null hypothesis. In other words, there is no evidence of a significant difference among the operators. See also CHI-SQUARED DISTRIBUTION.

F-test, *n.* **1.** a test, applied to the VARIANCES of two samples, of the NULL HYPOTHESIS, that they have both been drawn from the same parent population; **2.** a test, applied to the variance of a single sample, of the null hypothesis that it is drawn from a known normal parent population.

If two samples of sizes n_1 and n_2 are taken from the same normal parent population, then the estimates of the parent populations' variances, $\hat{\sigma}_1{}^2$ and $\hat{\sigma}_2{}^2$, should be approximately equal, particularly if n_1 and n_2 are large. If the ratio $\hat{\sigma}_1^2/\hat{\sigma}_2^2$ is not close to 1, then there may be reason to suspect that the two samples are drawn from different parent populations.

The F-distribution tables (pp. 250–253) list critical values for the ratio $\hat{\sigma}_1^2/\hat{\sigma}_2^2$ for different significance levels, for degrees of freedom $\nu_1 = n_1 - 1$ and $\nu_2 = n_2 - 1$. In applying the test, the sample with the greater variance is denoted number 1.

The *statistic F* is defined as

$$F = \frac{\hat{\sigma}_1^2}{\hat{\sigma}_2^2}$$

F has the *F-distribution*. If F is greater than the listed critical value, then the null hypothesis (that the samples are drawn from the same parent population) is rejected. This is a two-tail test (see ONE- AND TWO-TAIL TESTS), and so the percentage point to be looked up in the tables is half of the required significance level.

Example: two fishermen, Mr McTavish and Mr Desai, both catch some roach from a lake; the weights of their fish are as follows (in kg):

Mr McTavish	.15, .18, .25, .36, .42, .44	n=6
Mr Desai	.25, .26, .26, .30, .32, .33, .37, .37	n=8

Do the figures support, at the 10% significance level, the theory that they were fishing from the same shoal?

For Mr McTavish's fish, estimated parent variance $\hat{\sigma}^2 = .0154$.

For Mr Desai's fish, estimated parent variance $\hat{\sigma}^2 = .00234$.

So Mr McTavish's fish, having the greater variance, are called sample 1.

$$n_1 = 6 \qquad \hat{\sigma}_1^2 = .0154$$
$$n_2 = 8 \qquad \hat{\sigma}_2^2 = .00234$$
$$F = \frac{\hat{\sigma}_1^2}{\hat{\sigma}_2^2} = \frac{.0154}{.00234} = 6.58$$

The numbers of DEGREES OF FREEDOM are:

$$\nu_1 = 6 - 1 = 5 \qquad \nu_2 = 8 - 1 = 7$$

Since the F-test is two-tail, the percentage point to be looked up is $\frac{1}{2} \times 10\% = 5\%$. From the F-distribution tables (p. 250) for 5% points, the critical value for

$$F_7^5 = 3.97$$

Since $6.58 > 3.97$, the null hypothesis, that the samples came from the same parent population, is rejected. So the evidence suggests that, at the 10% significance level, the two men were not fishing the same shoal.

Since the F-test on the variances has failed, there is no point, or indeed validity, in applying a standard t-TEST on the differences in the sample means. The t-test assumes that both populations have a common variance.

If a sample is being compared with a known population, that population is treated as a sample with $\nu = \infty$. This is equivalent to a CHI-SQUARED (χ^2) TEST for variance.

function, *n.* a variable that can take a set of values, each of which is associated with the value of an independent variable or variables. Examples:
$P = 1000e^{.001t}$; P is a function of t
$y = 3 + \sin 2x$; y is a function of x
The notation $y = f(x)$ reads, y is a function of x.

A function must be unique for any value of its variable. Thus, $y = \text{inv} \sin (x)$ is not a function, since, for example, inv sin $(\frac{1}{2})$ can take values 30°, 150°, 390°. . .or

$$\frac{\pi}{6}, \frac{5\pi}{6}, \frac{13\pi}{6} \ldots \text{radians}$$

and so is not unique.

G

gamma distribution, *n.* the distribution with probability density function f(x) given by:

$$f(x) = \frac{x^{\alpha-1} e^{-\lambda x}}{\lambda^{-\alpha}\Gamma(\alpha)}$$

where α and λ are parameters of the distribution, and $\Gamma(\alpha)$ is the gamma function, given by:

$$\Gamma(\alpha) = \int_0^\infty x^{\alpha-1} e^{-x}\, dx \text{ (for } \alpha \text{ real and positive).}$$

If α is a positive integer, $\Gamma(\alpha) = (\alpha - 1)!$

The gamma distribution has

$$Mean = \frac{\alpha}{\lambda}$$

$$Variance = \frac{\alpha}{\lambda^2}$$

The graph of the gamma distribution varies in shape according to the value of α, as shown in Fig. 42.

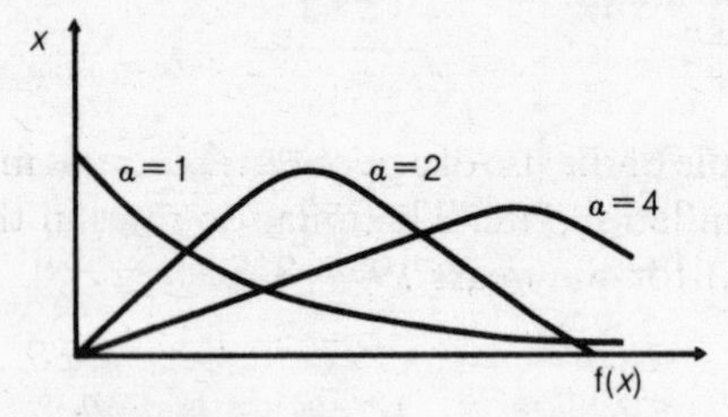

Fig. 42. **Gamma distribution.** The different curves for values 1, 2, 4 of the parameter α.

In the case $\alpha = 1$, the gamma distribution is the *exponential distribution.*

Gaussian curve, see NORMAL CURVE.

geometric distribution, *n.* the probability distribution $p(x)$ given by:

$$p(x) = pq^{x-1} \qquad \text{where } 0 < p < 1,\ q = 1 - p$$

It represents the probability of the first success in a series of BERNOULLI TRIALS occurring at the xth attempt. Each trial has probability p of success, probability q of $1 - p$ of failure.

For this distribution

$$Mean = \frac{1}{p}$$

$$Variance = \frac{q}{p^2}$$

Example: a die is thrown until a six occurs. What is the probability that the number of throws x is 5?

The distribution is geometric, with

$$p = \tfrac{1}{6} \qquad q = \tfrac{5}{6}$$

So the probability that $x = 5$ is:

$$\tfrac{1}{6} \times (\tfrac{5}{6})^4 = .08$$

Compare PASCAL'S DISTRIBUTION.

geometric mean, *n.* the average value of a set of n numbers, expressed as the nth root of their product. Thus, the geometric mean of n numbers $x_1, x_2 \ldots x_n$ is:

$$\sqrt[n]{x_1 x_2 \ldots x_n}$$

An example of the use of a geometric mean is in working out an average inflation rate. The inflation rates in the UK (retail prices index) for the years 1977–81 were:

1977	*1978*	*1979*	*1980*	*1981*
15.8%	8.3%	13.4%	18.0%	11.9%

The average inflation rate is that which, applied uniformly

over the same period, would have the same overall effect. To work this out, it is first necessary to see the inflation figures as scale factors by which the cost of living is increased, namely:

1977	*1978*	*1979*	*1980*	*1981*
1.158	1.083	1.134	1.180	1.119

Thus, goods which cost £1 at the start of 1977 cost £1 × 1.158p at the end of it, £1 × 1.58 × 1.083 at the end of the following year, and so on.

The overall effect of the five years is an increase by a scale factor of

$$1.158 \times 1.083 \times 1.134 \times 1.180 \times 1.119 = 1.878$$

corresponding to 87.8% over the five years.

A scale factor x per year, applied over the five years, would have the effect of x^5. Thus, for it to be equivalent,

$$x^5 = 1.878$$

$$x = \sqrt[5]{1.878} = 1.134$$

The average inflation rate was 13.4%.

The geometric mean may also be calculated by finding the arithmetic mean of the logarithms of the numbers, and then taking its antilogarithm. Thus, the above calculation could have been set out as:

Number	*Logarithm (base 10)*	
1.158	0.06371	
1.083	0.03463	
1.134	0.05461	
1.180	0.07188	
1.119	0.04883	
Total ÷ 5	0.27366	
	0.05473	Antilogarithm: 1.134

The average inflation rate was 13.4%.

goodness of fit, *n.* how well a particular set of data fits a given relationship or distribution. Goodness of fit may be measured, for grouped data, by the statistic χ^2 (see CHI-SQUARED TEST), where:

$$\chi^2 = \sum_{\text{all groups}} \frac{(O - E)^2}{E}$$

for the correct number of DEGREES OF FREEDOM, where O is the observed number in a group, and E the expected number.

grade, *n.* a position or degree in a scale, as of quality, rank, size, or progression.

Graeco-Latin square, see EXPERIMENTAL DESIGN.

graph, *n.* a drawing depicting the relationship between certain sets of numbers or quantities by means of a series of dots, lines, etc., plotted with reference to a set of axes.

It is important to realize that, when empirical data are plotted, the points should not necessarily be joined (Fig. 43a). Instead a line (or curve) of *best fit* may be drawn through them (as shown in Fig 43b).

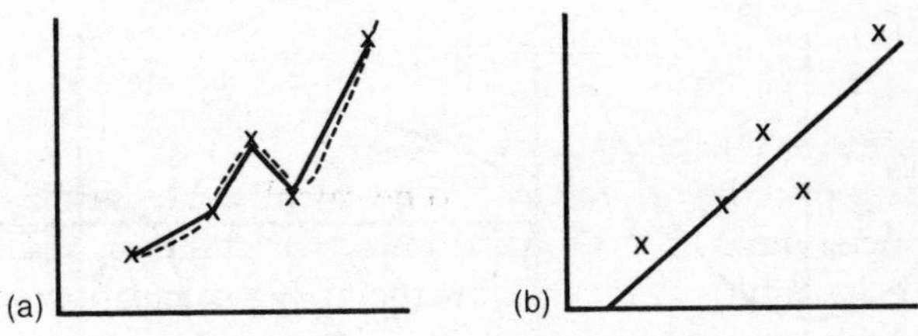

Fig. 43. **Graph.** (a) Data points incorrectly joined. (b) A line of best fit for the data shown in Fig. 43(a).

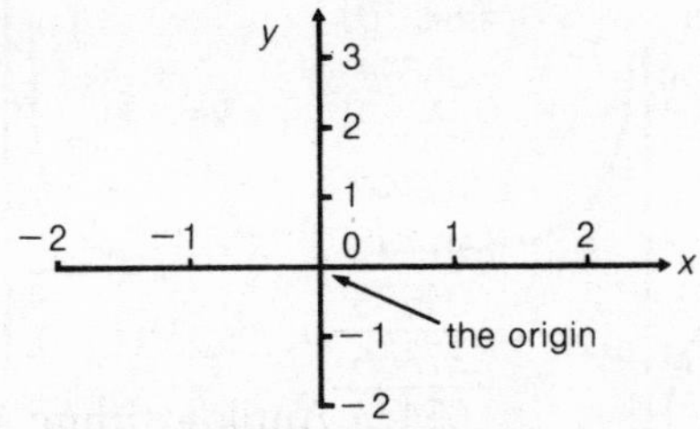

Fig. 44. **Graph.** Cartesian graph.

The most common type of graph is a *Cartesian graph,* with uniform scales on two perpendicular axes, often known as the *x*- and *y*-axes (Fig. 44). The *x*-axis is usually used for the independent variable, the *y*-axis for the DEPENDENT VARIABLE.

A particularly important Cartesian graph is the straight line, $y = mx + c$ (Fig. 45). This line has gradient m and intercept c, that is, it crosses the *y*-axis at the point (0, c). Other common Cartesian graphs are shown in Fig. 46.

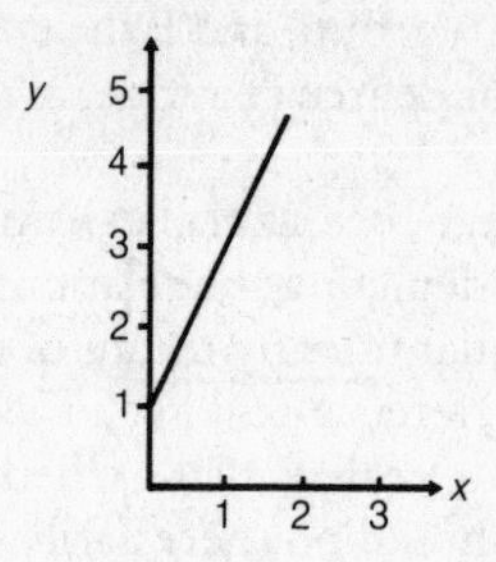

Fig. 45. **Graph.** The straight line, $y = mx + c$, illustrated by $y = 2x + 1$; the gradient m is 2, the intercept c is 1.

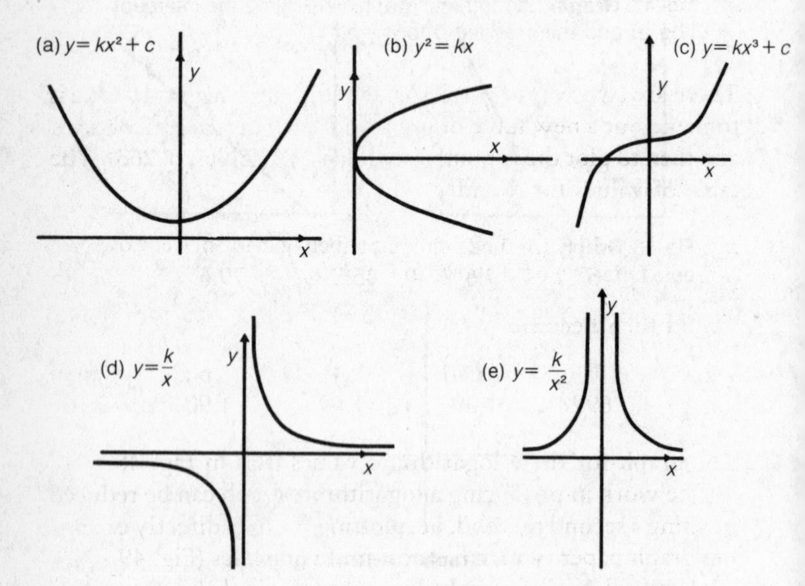

Fig. 46. **Graph.** Some common Cartesian graphs, where *k*, *c* are constants > 0. (a) $y = kx^2 + c$; (b) $y^2 = kx$; (c) $y = kx^3 + c$; (d) $y = k/x$; (e) $y = k/x^2$.

Logarithmic graphs. The relationship, $y=ax^b$, written in logarithmic form is:

$$\log y = b \log x + \log a$$

So, if log y is plotted against log x, a straight line should result (Fig. 47) with intercept log a and gradient b.

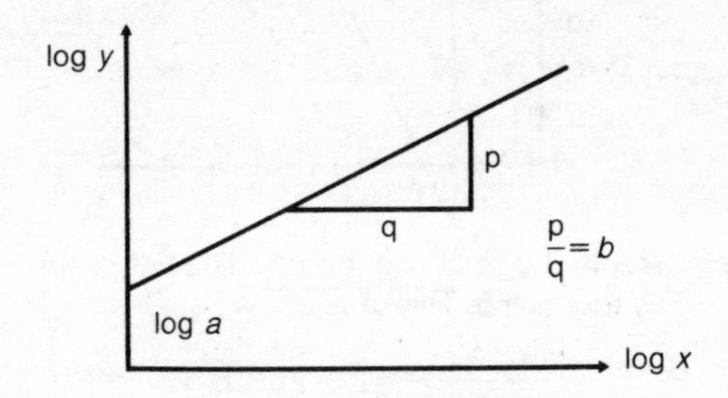

Fig. 47. **Graph.** Logarithmic graph, illustrating the intercept (log a) and the gradient b for $y=ax^b$.

There are two ways of arriving at a logarithmic graph. One is to make out a new table of log x and log y in place of x and y, and then to plot those points (see logarithm tables, p. 263). The table of values for x and y,

x	1	2	3	4	5
y	5	19.6	45.5	79.6	125

would then become

log x	0	.301	.477	.602	.699
log y	.699	1.29	1.66	1.90	2.10

The graph for these logarithmic values is as in Fig. 48.

The work in producing a logarithmic graph can be reduced by using a second method, i.e. plotting the data directly on log-log graph paper, which has non-uniform scales (Fig. 49). Not all data will give a straight line when plotted this way, only those points where the relationship is of the form $y=ax^b$, where a and b are constants.

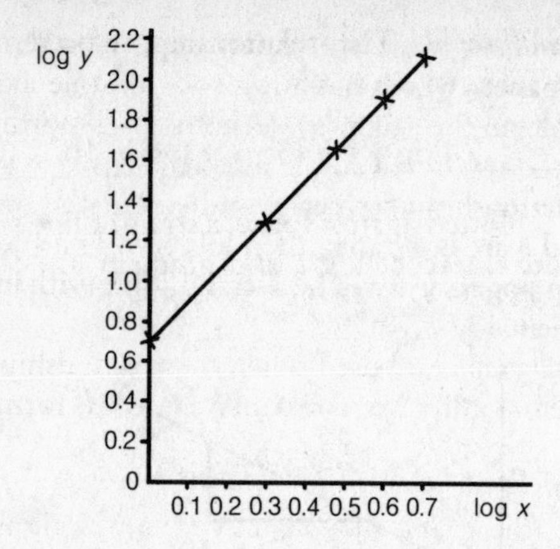

Fig. 48. **Graph.** Logarithmic graph. The intercept gives log a = .699; thus a = 5. The gradient is:

$$\frac{2.10-.699}{.699-0}=2.0;\text{ thus } b=2.$$

The relationship plotted is therefore $y=5x^2$.

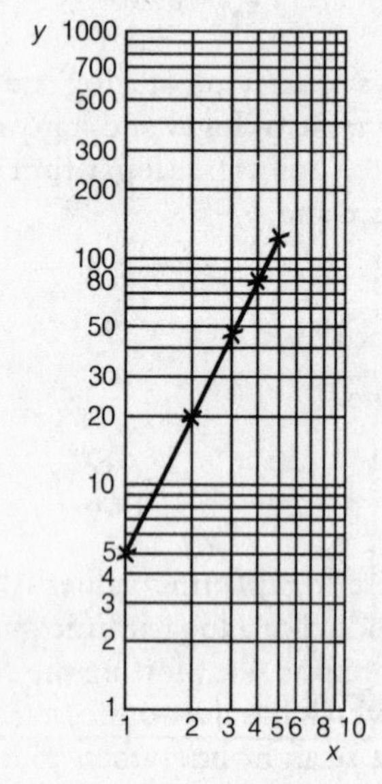

Fig. 49. **Graph.** The data used in Fig. 48 plotted on log-log graph paper.

Semi-logarithmic graphs show data plotted on semi-logarithmic graph paper, which has a log scale on one axis (y) and a uniform scale on the other (x). (Alternatively, ordinary graph paper may be used, and log y plotted against x.)

If the relationship between x and y is of the form $y = ab^x$, where a and b are constants, then $\log y = \log a + x \log b$. The graph of log y against x will be a straight line with intercept log a and gradient log b.

Square law graphs are used when the relationship between x and y (when a and b are constants) is of the form,

$$y = a + bx^2$$

In this case, the graph of y against x^2 is a straight line, with intercept a, gradient b. Data can be plotted directly on square law graph paper.

Example: the data below is plotted as a square law graph in Fig. 50.

x	1	2	3	4	5
y	5	11.2	20.6	35.3	52.1
x^2	1	4	9	16	25

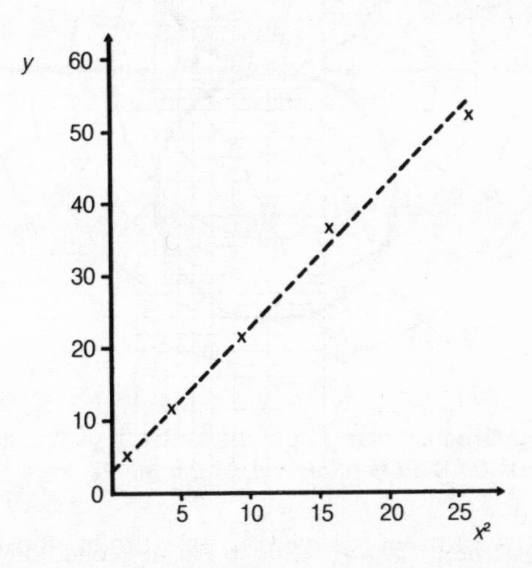

Fig. 50. **Graph.** Data plotted as a square law graph.

Polar graphs are centred on a particular point; the size of the variable depends on its direction from a fixed line through that point, usually taken to be in the same direction as the positive x-axis. Polar graph paper may be used.
Example: a simple radio aerial is used to pick up a signal from a distant source. The strength of the signal depends on the orientation of the aerial, as shown in the graph, Fig. 51.

Data which are thought to form a NORMAL DISTRIBUTION can be plotted on normal probability graph paper.

Several *charts* are published, from which probabilities can be read; they are a form of table. Examples include those used for POISSON DISTRIBUTION, BINOMIAL DISTRIBUTION, confidence intervals for correlation coefficients, etc.; see also FISHER'S Z-TRANSFORMATION.

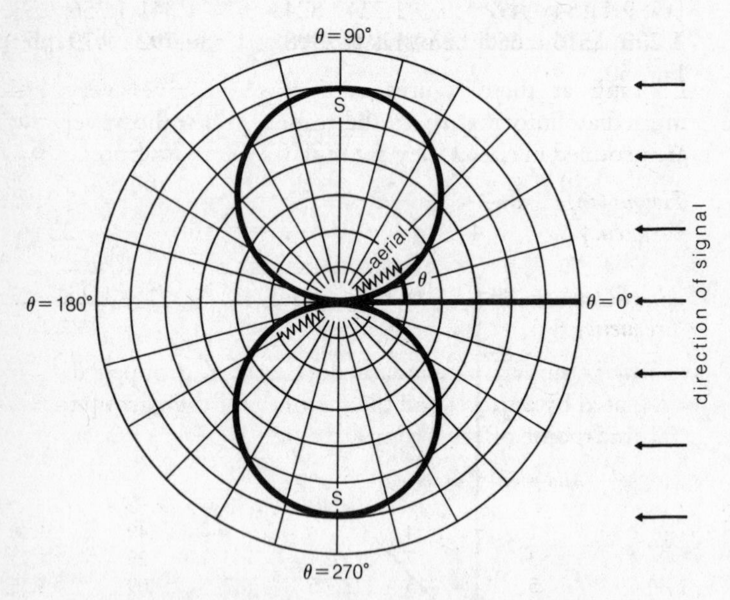

Fig. 51. **Graph.** Polar graph; the strength of the signal, S, depends on the orientation, θ, of the aerial.

group, *n.* a number of people or things considered as a collective unit.

grouped data, *n.* information which has been collected into groups or CLASSES for display, easy reading, or simplifying calculation.

Example: the heights of 100 waves were recorded one day at a coastal research station. They were as follows (in m):

.701	.192	1.002	.824	.702	.201	.889	.546	1.082	1.103
1.013	1.271	.212	.842	.936	.836	.402	.963	.614	.297
.303	.757	.768	.324	.959	.863	1.112	1.173	.812	1.054
.867	.942	.714	.723	1.521	.912	1.223	.625	1.145	.161
.746	.182	.646	1.216	.496	.934	.433	1.046	.419	.688
.736	.296	1.289	1.024	1.362	.946	1.292	.679	.836	.781
.573	.846	.584	.637	.991	.450	.476	.812	.369	.849
.836	.658	.742	1.039	.691	.823	1.082	.529	.962	1.330
.739	1.154	.492	.929	1.234	.824	.770	1.341	1.356	.538
1.256	.516	.669	.550	1.190	.998	.199	.792	.479	.901

Looking at these figures as they stand gives very little immediate information to the reader. When, however, they are grouped in classes they are much easier to interpret.

Height (m)	$0-<.2$	$.2\leqslant<.4$	$.4\leqslant<.6$	$.6\leqslant<.8$
Frequency	4	7	15	22

Height (m)	$.8\leqslant<1.0$	$1.0\leqslant<1.2$	$1.2\leqslant<1.4$	$1.4\leqslant<1.6$
Frequency	22	14	11	1

The mean $\bar{x}$ and standard deviation of grouped data are estimated by assuming all the members of each group to be at the mid-point of the group.

Height (m)	*Mid-point* x	*Frequency* f	xf	$(x-\bar{x})$	$(x-\bar{x})^2$	$(x-\bar{x})^2f$
0–.2	.1	4	.4	−.7	.49	1.96
.2–.4	.3	7	2.1	−.5	.25	1.75
.4–.6	.5	15	7.5	−.3	.09	1.35
.6–.8	.7	22	15.4	−.1	.01	0.22
.8–1.0	.9	26	23.4	.1	.01	0.26
1.0–1.2	1.1	14	15.4	.3	.09	1.26
1.2–1.4	1.3	11	14.3	.5	.25	2.75
1.4–1.6	1.5	1	1.5	.7	.49	.49
Total		100	80			10.04

$$\text{Mean } \bar{x} = \frac{\Sigma x\text{f}}{\Sigma \text{f}} = \frac{80}{100} = .8 \text{ (estimated)}$$

$$\text{Variance s}^2 = \frac{\Sigma (x-\bar{x})^2\text{f}}{\Sigma \text{f}} = \frac{10.04}{100} = .1004$$

$$\text{SD} = \sqrt{\text{Variance}} = .32$$

If the distribution has a central mode and tails off to either side, as in this case, this procedure will involve errors.

For groups below the mode, more of the members of a group will be above the mid-point than below it (Fig. 52). For groups above the mode, the opposite will be true.

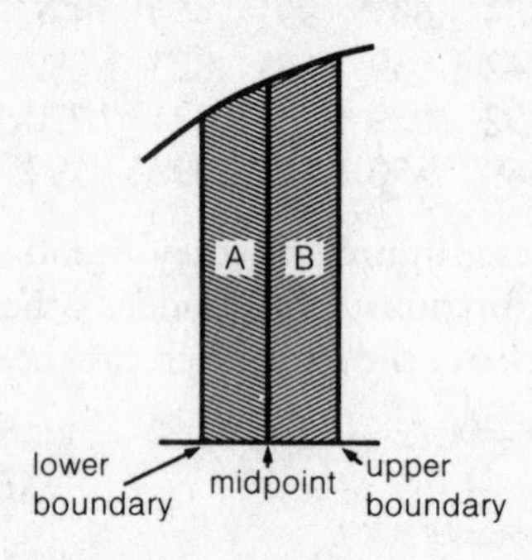

Fig. 52. **Grouped data.** The population B, above the midpoint of this class, is greater than that of A, below the midpoint. This is the case for any class below the mode.

For estimation of the mean, the errors above and below will approximately cancel each other out, so the method described is reasonably accurate. For standard deviation, however, the errors will add up so that the estimated value will be slightly too large (some of the population is nearer the middle than estimated). This can be allowed for by applying *Sheppard's correction*:

$$\textit{Corrected variance} = \text{calculated variance} - \frac{\text{c}^2}{12}$$

where c is the class interval.

For the figures in the example of wave heights:

$$\textit{Corrected variance} = .1004 - \frac{(.2)^2}{12} = .0971$$

$$\textit{Corrected standard deviation} = .312\text{m}$$

Sheppard's correction is only of value if the distribution is approximately normal. Thus it must be continuous (or discrete with no gaps in the data), and must tail off to zero either side of the mode. The class interval must be less than the standard deviation.

H

harmonic mean, *n.* the reciprocal of the ARITHMETIC MEAN of the reciprocals of a set of numbers. The harmonic mean m of the n numbers $x_1, x_2, \dots x_n$ is given by:

$$\frac{1}{m} = \frac{1}{n}\left\{\frac{1}{x_1} + \frac{1}{x_2} + \dots + \frac{1}{x_n}\right\}$$

Thus, the harmonic mean of 2, 4 and 5 is calculated:

$$\frac{1}{m} = \tfrac{1}{3}(\tfrac{1}{2} + \tfrac{1}{4} + \tfrac{1}{5}) \quad \text{giving } m = \tfrac{60}{19} = 3\tfrac{3}{19}$$

The harmonic mean is thus also given by:

$$\frac{n}{\sum_{i=1}^{n} \frac{1}{x_i}}$$

Example: a man travels 1 km at 5 km/h, 1km at 10km/h and 1km at 20km/h. What is his average speed?

$$\textit{Average speed} = \frac{\text{Distance travelled}}{\text{Time taken}}$$

$$= \frac{3\text{km}}{(\tfrac{1}{5} + \tfrac{1}{10} + \tfrac{1}{20})\text{ h}} = 8\tfrac{4}{7}\text{km/h}$$

Thus, the average speed is the harmonic mean of the three speeds.

heteroscedastic, see BIVARIATE DISTRIBUTION.

histogram, *n.* a chart for displaying grouped data in which the width of each bar is proportional to the class interval, and the area of each bar is proportional to the frequency it represents.

Example: in a survey of the village of Denstone, 126 householders were asked how long they had lived there. The results were grouped as follows.

Duration of residence (years)	0– <1	1⩽ <5	5⩽ <10	10⩽ <20	20⩽ <50
Frequency (number of householders)	6	39	30	27	24

Source: Denstone College Geography Department.

The histogram displaying this information is shown in Fig. 53.

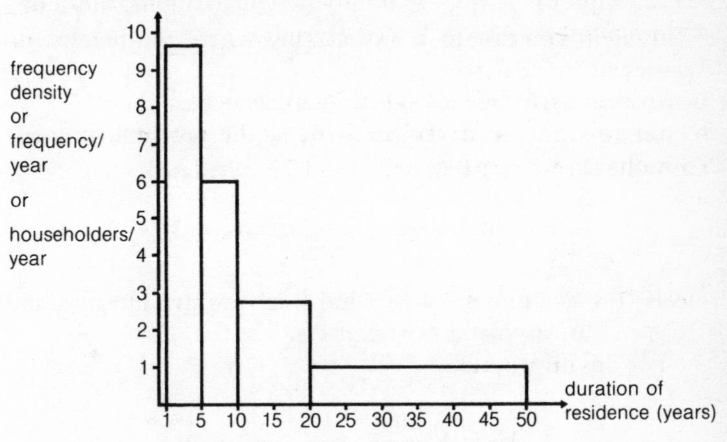

Fig. 53. **Histogram.** Results of a survey of householders in Denstone, of how long they had lived there.

(a) The heights of the bars are calculated by dividing the frequencies by the CLASS INTERVALS, measured in suitable units; thus, Householders ÷ years = Householders/year. In this case:

Class	*Frequency*	*Class interval*	*Height of bar (suitable unit)*
0⩽ <1	6	1	6
1⩽ <5	39	4	9.75
5⩽ <10	30	5	6
10⩽ <20	27	10	2.7
20⩽ <50	24	30	0.8

(b) the vertical scale is Frequency/year, or Householders/year, *not* frequency, or householders. This is sometimes written 'Frequency Density'.

(c) that it is the areas (not the heights) of the bars which give the frequencies. In this case:

$$(1-0)\ \times 6=6$$
$$(5-1)\ \times 9.75=39$$
$$(10-5)\ \times 6=30$$
$$(20-10)\times 2.7=27$$
$$(50-20)\times 0.8=24$$

This is in contrast to a BAR CHART, where the heights do represent frequencies.

homoscedastic, see BIVARIATE DISTRIBUTION.

hypergeometric distribution, *n.* the probability distribution p(x) is given by:

$$p(x)=\frac{{}^{a}C_{x}\,{}^{b}C_{n-x}}{{}^{a+b}C_{n}}$$

where $0\leqslant x\leqslant n$; $n\leqslant a+b$; a and b are positive integers, and ${}^{a}C_{x}$, etc., are BINOMIAL COEFFICIENTS.

For this distribution,

$$Mean=\frac{na}{a+b}$$

$$Variance=\frac{nab(a+b-n)}{(a+b)^{2}(a+b-1)}$$

This distribution is generated when a sample of size n is taken (without replacement) from a parent population containing a items of one particular type, and b of others. The probability that the sample contains exactly x of the particular type is given by the relevant term in the hypergeometric distribution.

hypothesis, *n.* a theory which is put forward either because it is believed to be true or because it is to be used as a basis for argument, but which has not been proved. Statistical hypotheses often refer to values of parent population PARAMETERS, like mean and variance. Setting up and testing hypotheses, particularly NULL HYPOTHESES, is an essential part of INFERENTIAL STATISTICS.

I

impossible, *adj.* incapable of occurring or happening. If an event is impossible, the PROBABILITY of its occurrence is zero.

increment, 1. *n.* a small change in the value of a variable. **2.** *vb.* to change the value of a variable systematically by small amounts.

independent events, *n.* events which have no influence on each other. If a die and a coin are thrown together, the events 'The die shows 6' and 'The coin comes heads' are independent.

If the knowledge that an event, A, has occurred influences the probability of another event, B, then B is not independent of A. For example, the event 'Mr Jones is wearing a raincoat' is unlikely to be independent of the event 'It is raining'.

This is expressed mathematically by saying that event A is independent of event B if

$$p(A|B) = p(A|B') = p(A)$$

In the same way, event B is independent of event A if

$$p(B|A) = p(B|A') = p(B)$$

In fact, if A is independent of B, then B must be independent of A and

$$p(A \cap B) = p(A)\ p(B)$$

See also CONDITIONAL PROBABILITY.

independent random variables, *n.* random variables, e.g. x and y, such that a knowledge of the value of x does not effect the probability distribution of y, and vice versa. Thus there is no relationship between the values of independent random variables, and their COVARIANCE is zero.

When two dice are thrown, their scores are independent; but the marks for papers I and II of a mathematics examination are not independent, many candidates performing well, moderately or badly on both papers.

This can be expressed more formally by saying that two random variables, X and Y, are independent if a knowledge of the value of one does not alter the probability distribution of the other. Thus

$$p(X=x \text{ and } Y=y)=p(X=x).\,p(Y=y)$$

for independent random variables, X and Y, taking values x and y.

independent variable, see DEPENDENT VARIABLE, EXPLANATORY VARIABLE.

index number, *n.* a statistic giving the value of a quantity (like the crime rate, or the cost of living) relative to its level at some fixed time or place which, conventionally, is given the number 100.

Example: in a certain country, the annual numbers of murders during the years 1976–80 were as follows:

1976	1977	1978	1979	1980
500	515	520	535	485

If 1976 is taken as the base year, 100, then the index number for 1977 is calculated as:

$$\frac{515}{500}\times 100=103$$

The full set of index numbers is thus

1976	1977	1978	1979	1980
100	103	104	107	97

In this example, the index was easily calculated because only one item, the number of murders, was under consideration. Index numbers which are used in practice, like the Index of Retail Prices, are usually built up from many items, and are calculated as a WEIGHTED MEAN of the indices for each of the items.

Example: a very simple cost-of-living index is constructed from three items, petrol, meat and potatoes, in the ratio 3:5:2. In the year taken as the base, and in the current year, prices were:

	Base year	*Current year*
Petrol (4l)	£1	£1.70
Meat (1kg)	£2.50	£2.80
Potatoes (25kg)	£1.50	£1.20

The individual index numbers are calculated as follows: The Base Year indices are all 100; the Current Year indices are

$$\text{Petrol} \quad \frac{1.70}{1} \times 100 = 170$$

$$\text{Meat} \quad \frac{2.80}{2.50} \times 100 = 112$$

$$\text{Potatoes} \quad \frac{1.20}{1.50} \times 100 = 80$$

The calculation is then continued in the table below

	Base index	*Current index number*	×	*Weighting*	=	*Current cost of living*
Petrol	100	170		3		510
Meat	100	112		5		560
Potatoes	100	80		2		160
Total				10		1,230

$$\text{Current index number} = \frac{1230}{10} = 123$$

The problem of the best weighting to give the different items in constructing an index number is one which has no correct answer. In the case of a cost of living index there might realistically be 600 component items. A correct weighting for one person need not be so for another; for example, for one man, beer may be a major source of expenditure while another, being teetotal, spends nothing at all on it. In practice, a

weighting scheme is determined on the basis of the average amount spent on each item. However, as tastes change with time, so does the ideal weighting; furthermore as new items become available and old ones obsolete, the list itself is liable to change.

In *Laspeyre's Index* the weighting is taken from the base year, in *Paasche's Index* from the current year. This means that Laspeyre's Index tends to emphasize items which are out of date and to ignore changes in taste or life style over the years. Paasche's index, on the other hand, runs into difficulties with new items which are significant in the current year but have no base value because they did not exist or were not used before. Imagine, for example, a cost of living index, set up with base year 1800 to compare the cost of living then with now. If Laspeyre's index is used, items like candles and horse feed will be quite heavily weighted, whereas electricity and petrol would not feature at all. If Paasche's index is used, the reverse is the case.

A compromise between the two is the *Typical Year Index*, where the weightings are taken for some typical time between the base year and the present.

In *Fisher's Ideal Index*, the value is taken as the GEOMETRIC MEAN of Laspeyre's Index and Paasche's Index. The *Marshall Edgeworth Bowley Index* takes the weighting for any item to be the average of that for the base year and that for the current year.

Formulae for the various indexes are as follows, where p_o = base year price, p_n = current year price, q_o = base year weighting, q_n = current year weighting, and q_b = typical year weighting:

$$\textit{Laspeyre's Index} = \frac{\sum p_n q_o}{\sum p_o q_o}$$

$$\textit{Paasche's Index} = \frac{\sum p_n q_n}{\sum p_o q_n}$$

$$\textit{Typical Year Index} = \frac{\sum p_n q_t}{\sum p_o q_t}$$

$$\textit{Fisher's Ideal Index} = \sqrt{\left\{\frac{\sum p_n q_o}{\sum p_o q_o}\right\}\left\{\frac{\sum p_n q_n}{\sum p_o q_n}\right\}}$$

$$\begin{array}{l}\textit{Marshall Edgeworth} \\ \textit{Bowley Index}\end{array} = \frac{\sum p_n(q_o + q_n)}{\sum p_o(q_o + q_n)}$$

inference, *n.* drawing conclusions about a PARENT POPULATION on the basis of evidence obtained from a SAMPLE.

inferential statistics, *n.* that branch of statistics which involves drawing conclusions from data by HYPOTHESIS testing. Compare DESCRIPTIVE STATISTICS. See also INTERPRETIVE STATISTICS.

infinite, *adj.* having no limits or boundaries (of time, space, extent or magnitude).

infinitesimal, *adj.* **1.** infinitely or immeasurably small. **2.** of, relating to, or involving, a small change in the value of a variable that approaches zero as a limit.

integer, *n.* a whole number; 5, −17 and 0 are all integers.

intelligence quotient, IQ, *n.* a measure of the intelligence of an individual. Intelligence quotient is given by:

$$IQ = \frac{\text{Mental age}}{\text{Chronological age}} \times 100$$

Intelligence quotient has mean value 100, standard deviation 15.

interaction, *n.* when data is grouped according to several classifications which are not independent, there is said to be interaction between the classifications.

interpolation, *n.* estimation of a value of a variable between two known values. Compare EXTRAPOLATION.

Example: Alexandra was 1.21m tall on 1 January 1982 and 1.25m tall a year later, on 1 January 1983. Estimate her height on 1 April 1982.

This could be found graphically or by calculation.

$$\begin{aligned}\text{1 Jan to 1 April} &= 3 \text{ months} \\ \text{One year} &= 12 \text{ months}\end{aligned}$$

So her height on 1 April was:

$$\frac{3}{12} \times (1.25 - 1.21) + 1.21 = 1.22\text{m}$$

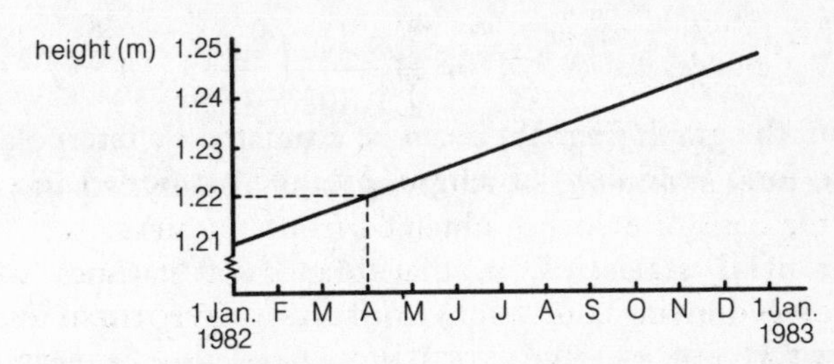

Fig. 54. **Interpolation.** From the actual values plotted for Alexandra's height on 1 January 1982 and 1 January 1983, her height on 1 April 1982 can be interpolated (1.22m).

From the graph (Fig. 54) it can be estimated by interpolation that Alexandra's height on 1 April was 1.22m. This answer assumes that the graph of Alexandra's height plotted against time was a straight line, indicating that she grew uniformly over the year. This may very well not have been the case; she could, for example, have grown more in the second half of the year than in the first, or vice versa. However, in the absence of any other information, it is usual to assume uniform change; the method is then called *linear interpolation*.

It is common to use linear interpolation when reading from tables. The figures tabulated here come from the table for the height of the normal curve (see normal probability tables, p. 246):

x	1.0	1.1
$\varphi(x)$	.242	.218

The value of $\varphi(1.06)$, found by linear interpolation to be .228, is arrived at as follows:

$$.242 + \frac{(1.06 - 1.00)}{(1.10 - 1.00)} \times (.218 - .242) = .228$$

Non-linear interpolation is sometimes used when enough points are known to draw a curved graph.

Example: in an experiment on a simple pendulum, a student found values of the period T for the different pendulum lengths l.

$l(m)$	.1	.2	.4	.6	.8	1.0
$T(s)$	.6	.9	1.25	1.55	1.8	2.0

From the graph (Fig. 55) it can be estimated by interpolation that, for a pendulum of length .3m, the period is 1.08s.

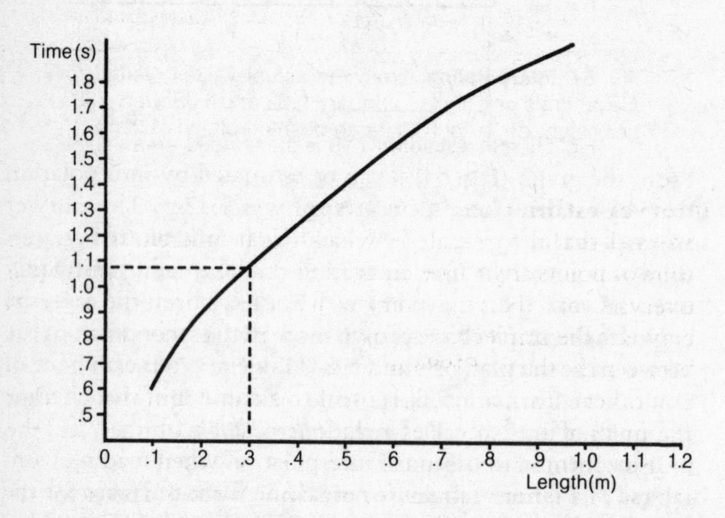

Fig. 55. **Interpolation.** Graph illustrating non-linear interpolation.

interpretive statistics, *n.* that branch of statistics concerned with drawing conclusions from data. Compare DESCRIPTIVE STATISTICS.

interquartile range, *n.* the difference between the upper and lower QUARTILES of a set of numbers. For example, the numbers

2 3 4 5 6 6 6 7 7 8 9

have upper quartile 7, lower quartile 4; the interquartile range is thus $7-4=3$.

Half of the interquartile range (in this example, $\frac{1}{2} \times 3 = 1\frac{1}{2}$) is called the *semi-interquartile range* (or *quartile deviation*) and is a measure of DISPERSION or spread comparable with standard deviation and mean absolute deviation.

intersection, *n.* the elements that are common to two (or more) SETS (Fig. 56). The intersection of sets A and B is written $A \cap B$. For example, if A = {s, t, a, r} and B = {p, l, a, n, e, t}, then $A \cap B$ = {a, t}.

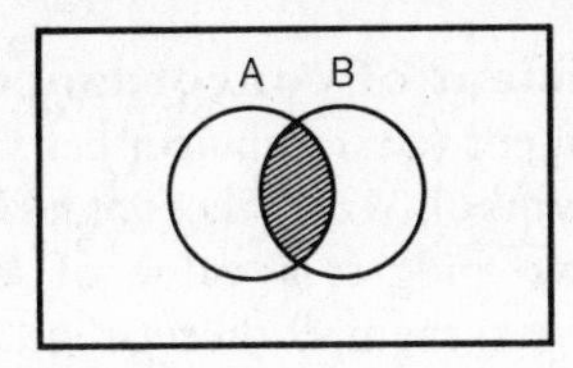

Fig. 56. **Intersection.** $A \cap B$ is the shaded area.

interval estimation, see POINT ESTIMATION.

interval scale, *n.* a scale in which equal differences between pairs of points anywhere on the scale are represented by equal intervals, but the zero point is arbitrary. The time interval between the starts of years 1981 and 1982 is the same as that between the starts of 1983 and 1984, namely 365 days. The zero point, year 1 AD, is arbitrary; time did not begin then. Other examples of interval scales include the heights of tides, and the measurement of longitude.

inverse Fisher transformation, see FISHER'S Z-TRANSFORMATION.

irregular variation, *n.* that variation within a TIME SERIES which is not accounted for by TREND, SEASONAL COMPONENTS or CYCLIC COMPONENTS.

K

Kendall's coefficient of concordance, *n.* a measure of the level of agreement (concordance) between several sets of matched results. Symbol: W. It takes values between 0 and 1. If there are only two judges, a value of W near O means disagreement between them; if there are many, a value near 0 means randomness in their judgments. A value near 1 shows agreement between the judges. See also CORRELATION.
Example: in testing a new range of cakes, labelled A to F, a bakery asks four people to rank them in order of preference. The results are:

	A	*B*	*C*	*D*	*E*	*F*	
Mrs McDiarmid	1	6	2	5	3	4	
Mrs Nandi	2	5	1	3	4	6	
Mrs O'Leary	1	4	3	6	5	2	
Mr Patrice	2	4	3	6	1	5	
Sum of ranks R_i	6	19	9	20	13	17	$=84$
Mean of ranks $\bar{R}$	$\frac{84}{6}=14$						
$(R_i-\bar{R})$	-8	5	-5	6	-1	3	
$(R_i-\bar{R})^2$	64	25	25	36	1	9	
Sum S	64 +	25 +	25 +	36 +	1 +	9	$=160$

The first steps in the calculation are shown under the table of results. The ranks are summed for each of the six cakes, giving the figures listed as R_i. The average of these, $\bar{R}$, is calculated as:

$$\tfrac{1}{6}(6+19+9+20+13+17)=14$$

The next two rows give the values of $(R_i-\bar{R})$ and $(R_i-\bar{R})^2$.

The values of $(R_i - \bar{R})^2$ are summed to give the quantity S, in this case 160.
Kendall's coefficient, W, is then calculated using the formula:

$$W = \frac{12S}{K^2(N^3 - N)}$$

where $S = \sum_{i=1}^{N} (R_i - \bar{R})^2$ (which is 160 in this example) K is the number of judges (4 in this case), and N is the number of objects judged (6 in this case). Kendall's coefficient of rank concordance, W, is thus:

$$W = \frac{12 \times 160}{16(216 - 6)} = .57$$

Kendall's coefficient of rank correlation, see CORRELATION COEFFICIENT.

Kruskal-Wallis one-way analysis of variance, *n.* a non-parametric test of the NULL HYPOTHESIS that three or more samples are drawn from the same PARENT POPULATION. Symbol: H. The test is carried out on the overall ranks of the items sampled, and so is a test for differences in location of the particular samples.
Example: 21 expectant mothers followed three different diets, A, B and C. At birth, the weights of their babies (in kg) were as follows:

Diet A	*Diet B*	*Diet C*
2.8	3.4	2.9
3.5	3.2	3.9
2.7	3.3	3.1
3.7	3.3	3.9
3.7	3.6	4.0
3.8	3.0	4.1
2.8		4.2
		4.0

Can it be claimed, at the 5% significance level, that the different diets had an effect on the weight of the babies?

Null hypothesis: There is no difference between the birth-weights of babies whose mothers have been on the different diets.

The babies are ranked from 1 to 21, according to weight, and then the statistic H is calculated, where

$$H=\frac{12}{N(N+1)}\left(\sum\frac{Rj^2}{Nj}\right)-3(N+1)$$

In this case there are three groups, so j takes the values 1, 2, and 3.

N_1, N_2, N_3 are the sizes of the three different groups, totalling N; $N_1=7$, $N_2=6$, $N_3=8$; N is thus $7+6+8=21$. R_1, R_2, R_3 are the sums of the overall ranks for each group, and are calculated from the table of ranks for each baby, below.

	A	*B*	*C*
	19.5	12	18
	11	15	5.5
	21	13.5	16
	8.5	13.5	5.5
	8.5	10	3.5
	7	17	2
	19.5		1
			3.5
Total of ranks Rj	95	81	55
Rj²	9,025	6,561	3,025

$$H=\frac{12}{21\times(21+1)}\left(\frac{9{,}025}{7}+\frac{6{,}561}{6}+\frac{3{,}025}{8}\right)-3(21+1)=5.71$$

H is then compared with the critical value of χ^2 for $K-1$ DEGREES OF FREEDOM, K being the number of groups. In this example, at the 5% level for 2 degrees of freedom, the CRITICAL VALUE for χ^2 (from the table on p. 249) is 5.99 (see CHI-SQUARED DISTRIBUTION).

Since $5.71<5.99$, the null hypothesis cannot be rejected at this level. Thus, there is no evidence, at the 5% level, on which to claim significant differences in the weights of babies whose mothers used the three diets. It should, however, be noted that the value of H is close to the critical value. In such a situation it is often desirable to carry out more tests to clarify the position.

kurtosis, *n.* the sharpness of a peak on a curve of a probability density function (see Fig. 57). A distribution with high kurtosis is called *leptokurtic*; one with medium kurtosis, *mesokurtic*, and one with low kurtosis, *platykurtic*.

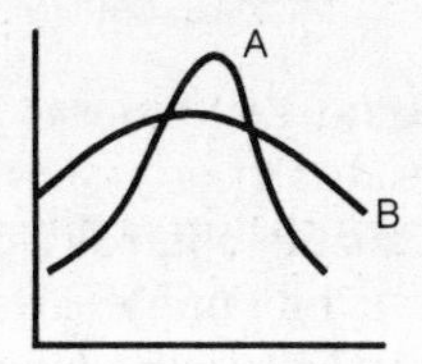

Fig. 57. **Kurtosis.** Distribution A has larger kurtosis than B.

A measure of kurtosis is the *coefficient of kurtosis*, which is defined as:

$$\frac{\mathrm{E}[(X-\mu)^4]}{\sigma^4} - 3$$

See also MOMENT.

L

Laspeyre's index, see INDEX NUMBER.

Latin square, see EXPERIMENTAL DESIGN.

lattice diagram, *n.* in quality control, a diagram on which the results of ATTRIBUTE TESTS on samples are illustrated. Example: the first six items to be tested are Good, Good, Defective, Good, Defective, Good. Starting at the bottom left of a lattice diagram (Fig. 58), a line is drawn horizontally to the right every time an item is good, vertically up if it is bad. The bent line which results is called the *sampling line*.

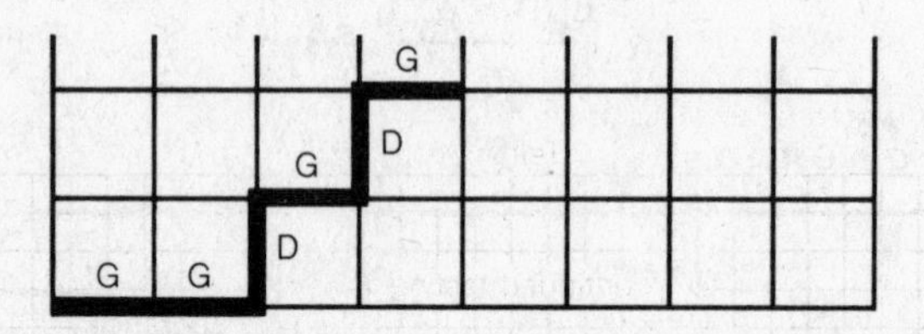

Fig. 58. **Lattice diagram.** This sampling line illustrates whether the first six items to be tested were good G or defective D in a quality-control test.

In a DOUBLE SAMPLING SCHEME, one sample is taken; if the number of defective items is not more than the ACCEPTANCE NUMBER, A_1, the whole batch is accepted. If it is greater than another larger number, A_2, the batch is rejected or subjected to 100% inspection. If the number of defectives lies between A_1 and A_2, a second sample is taken. If, after the second sample is taken, the total number of defectives is greater than a new acceptance number, A_3, the whole batch is rejected or subjected to 100% inspection.

First sample: Size = 40 $A_1 = 3$ $A_2 = 8$
By the end of the first sample, the batch will have been rejected if the number of defectives is over 8, the upper line on the lattice in Fig. 59. It will have been accepted if the outcome is one of:

(40 G, 0 D)
(39 G, 1 D)
(38 G, 2 D)
(37 G, 3 D).

These points form the lower line of the lattice diagram. Thus, these two lines form boundaries for the three regions, *acceptance* (bottom right), *rejection* (top) and *further testing* (middle).

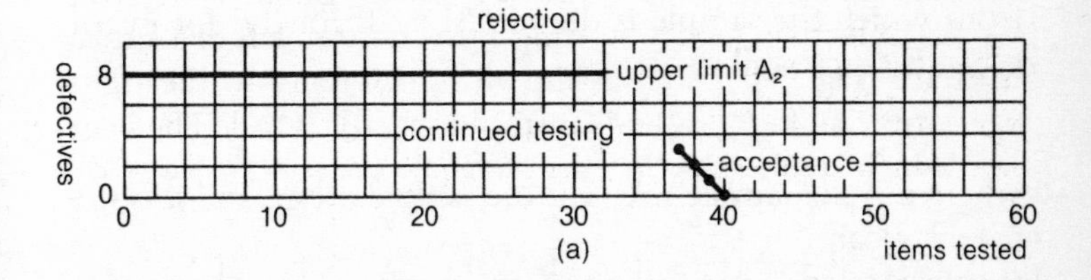

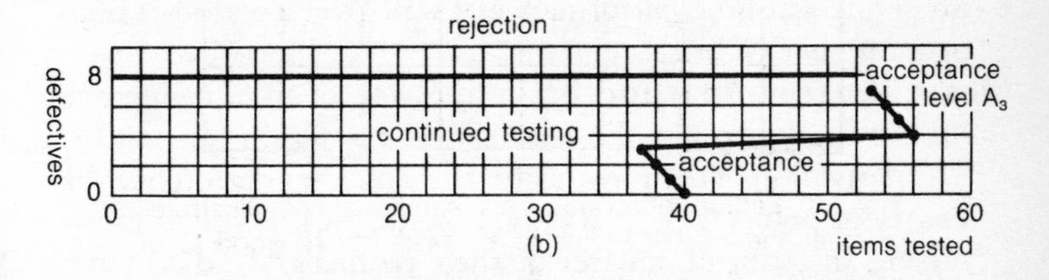

Fig. 59. **Lattice diagram.** (a). After a first sample batch of 40 items. (b). Showing the results of a second sample batch of 20 items.

A second sample batch is then tested, and a new acceptance number A_3 set (usually at the level of A_2 from the first sample, or higher).
Second sample: Size = 20 $A_3 = 8$
It will have been accepted if the total outcome of sampling is one of:

(53 G, 7 D)
(54 G, 6 D)
(55 G, 5 D)
(56 G, 4 D)

The results for the second sample are plotted on the same lattice diagram, Fig. 59b.

The idea of having three regions is not confined to double sampling schemes. In a SEQUENTIAL SAMPLE, the lower line goes up in a series of steps, and sampling is continued until the sampling line for the batch enters either the acceptance or the rejection region.

law of large numbers, *n.* the law which states that the larger a sample, the nearer its mean is to that of the parent population from which the sample is drawn. More formally: for every $\epsilon > 0$, the probability

$$\{|\bar{x} - \mu| > \epsilon\} \to 0 \quad \text{as } n \to \infty$$

where n is the sample size, $\bar{x}$ is the sample mean, and μ is the parent mean.

learning curve, *n.* a graph of the results of an experiment involving learning, performance or skill, over a period of time or a number of tests.

least squares, method of, *n.* the way of minimizing the sum of the squares of the residuals, as a criterion for best fit. The derivation of the formula for the y on x regression line for the set of points $(x_1, y_1), (x_2, y_2), \ldots, (x_n, y_n)$, involves minimizing the sum of squares of their residuals, $d_1, d_2, \ldots, d_n$. Thus, if the regression line is taken to be $y = mx + c$ (Fig. 60) the residuals are $y_1 - (mx_1 + c)$, $y_2 - (mx_2 + c)$, ..., $y_n - (mx_n + c)$.

So the sum of the squares of the residuals is given by:

$$S = \sum_{i=1}^{n} d_i^2 = \sum_{i=1}^{n} [y_i - (mx_i + c)]^2$$

This is then minimized by setting $\frac{\partial S}{\partial m}$ and $\frac{\partial S}{\partial c}$ both $= 0$.

This gives a pair of simultaneous equations in m and c,

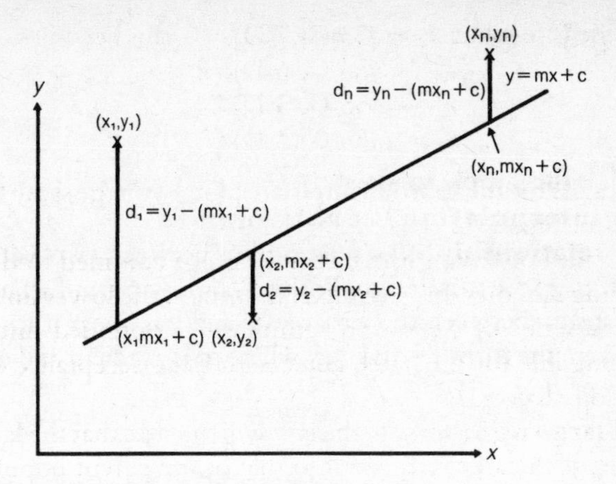

Fig. 60. **Least squares, method of.** Residuals from y on x regression line.

which are solved to give:

$$m = \frac{s_{xy}}{s_x^2} \qquad c = \bar{y} - \frac{s_{xy}}{s_x^2}\bar{x}$$

($\bar{x}$, $\bar{y}$ are the mean values of x and y, s_x the standard deviation of x, and s_{xy} the covariance of x and y.)

The regression line is thus:

$$y - \bar{y} = \frac{s_{xy}}{s_x^2}(x - \bar{x})$$

While the method of least squares is of theoretical importance, it is not often used directly as it is incorporated in formulae like that derived above for the y on x regression line.

linear estimator, *n.* an ESTIMATOR which is a linear function of the sample values. A linear estimator for the parent mean is:

$$\frac{a_1x_1 + a_2x_2 + \dots + a_nx_n}{a_1 + a_2 + \dots + a_n}$$

where $x_1, x_2, \dots, x_n$ are the sample values and $a_1, a_2, \dots, a_n$ all constants.

In the case when $a_1 = a_2 = \ldots = a_n = 1$, this becomes

$$\frac{x_1 + x_2 + \ldots + x_n}{n}$$

which is the sample MEAN, $\bar{x}$.

linear interpolation, see INTERPOLATION.

linear relationship, *n.* a relationship between two variables which gives rise to a straight line on a Cartesian graph. Thus the relationship between the variables x and y is linear if it can be written in the form $y = mx + c$, where m is gradient and c is the intercept (Fig. 61).

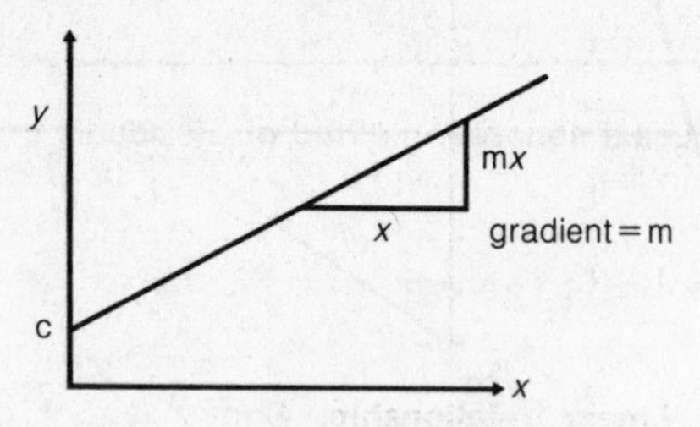

Fig. 61. **Linear relationship.**

Example: the cost of printing n books is £C where C = 2n + 5000. This is a linear relationship. £5000 is the cost of setting up the presses etc., and £2 the cost of printing each book thereafter.

It is sometimes possible to convert a non-linear relationship

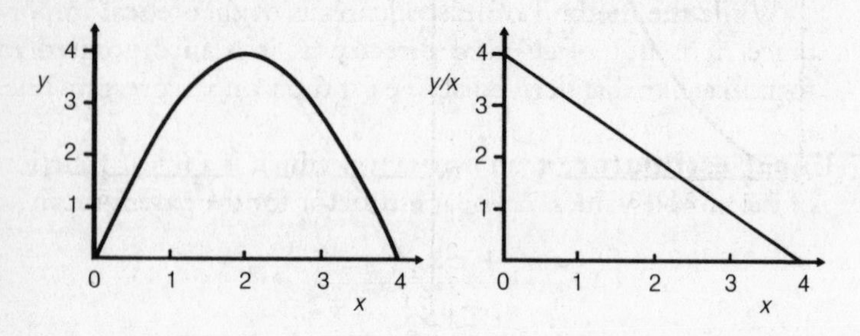

Fig. 62. **Linear relationship.** Conversion of non-linear relationship; $y = 4x - x^2$ becomes $y/x = 4 - x$.

into a linear one, by algebraic manipulation and a change of variables. There are two major advantages in graphing a relationship in linear form. One is that it allows the equation to be worked from the intercept and gradient of the graph, the other that OUTLIERS are easily spotted. Examples are shown in Figs. 62, 63 and 64.

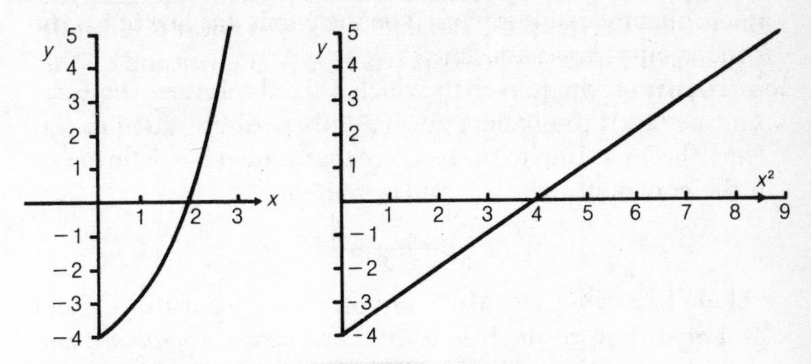

Fig. 63. **Linear relationship.** Conversion of non-linear relationship; $y = -4 + x^2$ is plotted as y against x^2.

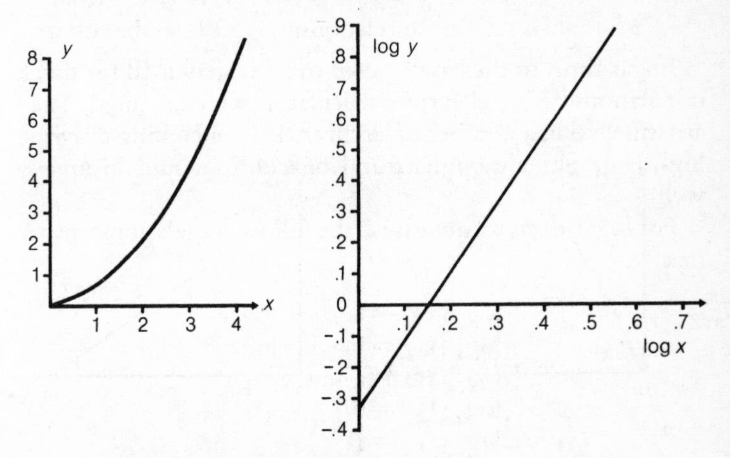

Fig. 64. **Linear relationship.** Conversion of non-linear relationship; $y = \frac{1}{2}x^2$ becomes $\log y = 2\log x + \log \frac{1}{2}$.

line of best fit, *n.* the line which fits a set of data most accurately when plotted on a graph. It depends on how greatest accuracy is defined in any example. Usually the least squares definition is used (see LEAST SQUARES, METHOD OF), so that the sum of the squares of the RESIDUALS should be a minimum. Usually the independent variable (see DEPENDENT VARIABLE) is plotted on the x-axis, being known accurately; the least squares line of best fit is then the y on x REGRESSION LINE. (If the independent variable is plotted on the y-axis, the line of best fit is the x on y regression line.)

logarithm, *n.* the power to which a fixed number, the base, must be raised to obtain a given number. Abbreviated to *log*. Thus, the logarithm to the base a of the number b is defined as c in the equation $a^c = b$. and is written as:

$$c = \log_a b$$

Two bases are particularly important for logarithms, e and 10. Logarithms to the base e are called *natural logarithms*, and often denoted by ln; ($\log_e$ is also correct). Natural logarithms arise because:

$$\int_1^X \frac{1}{x} dx = \ln X$$

Logarithms to the base 10 used to be widely used for quick calculations before electronic calculators were invented. They are still used in a number of circumstances, including drawing log-log graphs (although natural logarithms would do equally well).

For logarithms to any base a, the following relationships are true:

$$\begin{aligned} \log_a (xy) &= \log_a x + \log_a y \\ \log_a (x/y) &= \log_a x - \log_a y \\ \log_a (x^n) &= n \log_a x \\ \log_a (1) &= 0 \\ \log_a (a) &= 1 \end{aligned}$$

for $x > 0$, $y > 0$, $a > 1$.

All logarithmic graphs have the same shape, illustrated in Fig. 65 for base a (a > 1).

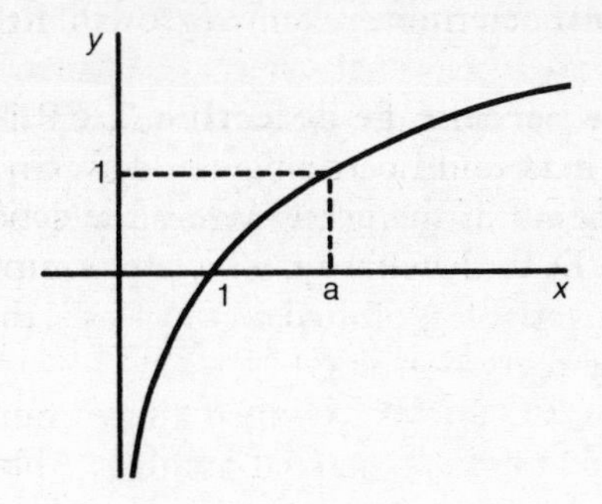

Fig. 65. **Logarithm.** Graph of $y = \log_a x$.

It is not possible to have the logarithm of a negative number. As $x \to 0$, $\log_a x \to -\infty$. As $x \to \infty$, $\log_a x \to \infty$.

The inverse of a logarithm is called its *antilogarithm* (antilog). If $y = \log_a x$, then $x = \text{antilog}_a y$. Thus $\text{antilog}_a y = a^y$.

logarithmic graph, see GRAPH.

logistic curve, *n.* a curve which is often used to model the growth of a variable. Its equation is of the form:

$$f(x) = a/(1 + be^{-cx})$$

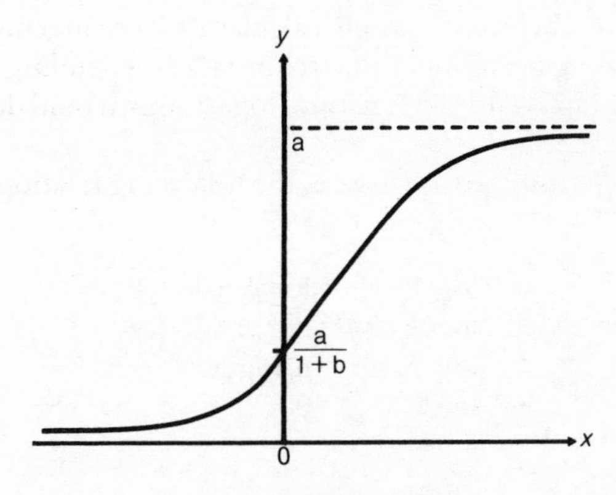

Fig. 66. **Logistic curve.**

The values of *a*, *b* and *c* are assigned for the particular variable under consideration; *a* represents the ultimate magnitude of the variable, *b* and *c* determine its rate of growth. Its graph is shown in Fig. 66.

lot tolerance percentage defective, LTPD, *n.* (in quality control) the maximum percentage of defectives accepted in a sampling scheme. If the percentage of defectives is no more than the LTPD, the batch from which the sample was drawn is accepted.

M

Malthus' theory, *n.* the theory that the human population will expand exponentially while the means of subsistence will increase more slowly, with the consequence that ultimately its size will be kept in check by famine, disease and war. It was put forward by the English economist T.R. Malthus (1766–1834). The widespread use of birth control methods during this century has made much smaller rates of population growth possible; consequently the theory is now mainly of historical interest.

Mann Whitney U-test, *n.* a non-parametric test of the NULL HYPOTHESIS that two samples are drawn from the same parent population, using their combined rank orders.

Example: two comparable groups of students were taught the same syllabus by different teachers. At the end of the course they took the same examination, gaining these marks:

Mr Leroy's group	65	42	39	58	61	72		
Mrs French's group	25	41	63	56	47	49	28	45

Is there evidence at the 5% significance level that one teacher is better than the other?

Null hypothesis: There is no difference between the groups. Values of the statistics U_1 and U_2 are calculated from the formulae,

$$U_1 = n_1 n_2 + \frac{n_1(n_1+1)}{2} - R_1 \qquad U_2 = n_1 n_2 + \frac{n_2(n_2+1)}{2} - R_2$$

where n_1 and n_2 are the sizes of the two groups, R_1 and R_2 the sums of the ranks of their members; $n_1 = 6$, $n_2 = 8$ in this case. In this example the ranks are as follows:

Mr Leroy	2	10	12	5	4	1		
Mrs French	14	11	3	6	8	7	13	9

giving

$$R_1 = 2 + 10 + 12 + 5 + 4 + 1 = 34$$
$$R_2 = 14 + 11 + 3 + 6 + 8 + 7 + 13 + 9 = 71$$

so that

$$U_1 = 6 \times 8 + \frac{6 \times 7}{2} - 34 = 35$$

$$U_2 = 6 \times 8 + \frac{8 \times 9}{2} - 71 = 13$$

(A useful check at this stage is that $U_1 + U_2$ should equal $n_1 n_2$; thus, $35 + 13 = 6 \times 8 = 48$.)

The values of U_1 and U_2 are then checked against the tables for the U-test (see p. 258), in this case a two-tail test (see ONE- AND TWO-TAIL TEST). For $n_1 = 6$, $n_2 = 8$, at 5% significance level, the tables give critical values of U of 8 and 40. So, for rejection of the null hypothesis, one value of U must be less than 8, the other greater than 40. In this case,

$$U_1 = 35 < 40$$
$$U_2 = 13 > 8$$

So there are no grounds for rejecting the null hypothesis. There is not evidence, at the 5% level, to claim either teacher was better than the other.

If the value of n_1 or n_2 (or both) is too large to allow the tables to be used (i.e. > 20), the test is carried out using the smaller of U_1 and U_2, calling it U.

On the null hypothesis of no difference between the two groups, U is distributed approximately normally with:

$$\textit{Mean} = \tfrac{1}{2} n_1 n_2$$
$$\textit{Standard deviation} = \sqrt{\frac{n_1 n_2 (n_1 + n_2 + 1)}{12}}$$

When ranks are tied the standard deviation is given by:

$$SD=\sqrt{\frac{n_1n_2}{N(N-1)}\left(\frac{N^3-N}{12}-\sum_i T_i\right)}$$

where $n_1+n_2=N$, and,

$$T_i=\frac{t_i^3-t_i}{12}$$

where t_i is the number of observations tied at rank i.

marginal distribution, see BIVARIATE DISTRIBUTION.

Markov chain, *n.* a sequence of events, the probability of each of which is dependent on the event immediately preceding it, but independent of earlier events. (Named after A. Markov, 1856–1922, Russian mathematician). Every Markov chain has a TRANSITION MATRIX. In certain cases, it is also a STOCHASTIC PROCESS.

Marshall Edgeworth Bowley Index, see INDEX NUMBER.

matched groups, *n.* groups which are matched for similarity before an experiment is carried out on one of them, the other being used as the control group (see CONTROL EXPERIMENT).

matched sample, *n.* **1.** a sample in which the same attribute, or variable, is measured twice, under different circumstances. **2.** two samples in which the members are clearly paired; see PAIRED SAMPLE.

mean, *n.* **1.** Average; usually arithmetic mean, unless otherwise stated.

There are several types of mean, used in appropriate circumstances. For the numbers $x_1, x_2, \ldots, x_n$:

ARITHMETIC MEAN $\quad \dfrac{x_1+x_2+\ldots+x_n}{n}$

GEOMETRIC MEAN $\quad \sqrt[n]{x_1x_2\ldots x_n}$

HARMONIC MEAN $\quad \dfrac{1}{\frac{1}{n}\left(\frac{1}{x_1}+\frac{1}{x_2}+\ldots+\frac{1}{x_n}\right)}$

WEIGHTED MEAN $\dfrac{w_1x_1 + w_2x_2 + \ldots w_nx_n}{w_1 + w_2 + \ldots + w_n}$

where w_1, w_2, ... w_n are the weightings of the numbers x_1, $x_2 \ldots x_n$.

2. Expectation, the expected value of a variable, as given by:

$$E(X) = \sum_i x_i p(x_i)$$

for a DISCRETE VARIABLE, and

$$E(X) = \int x f(x)\, dx$$

for a CONTINUOUS VARIABLE.
If $g(X)$ is a function of the random variable X, then

$$E(g(X)) = \int g(x) f(x)\, dx.$$

For example,

$$E(X^m) = \int x^m f(x)\, dx$$
$$E(e^X) = \int e^x f(x)\, dx$$

where $f(x)$ is the probability density function of X.

mean absolute deviation, *n.* the mean of the absolute values of the differences between the values of a variable and the mean of its distribution. See DISPERSION. Compare STANDARD DEVIATION, INTERQUARTILE RANGE.

measures of central tendency, *n.* typical or central values of a variable. Commonly used measures of central tendency are MEAN (or average), MODE, MEDIAN and MIDRANGE.

measures of dispersion, see DISPERSION.

median, *n.* the middle value in a distribution, below and above which lie values with equal total frequencies or probabilities. Thus, if there are n values, the median is that ranked $(n + 1)/2$. Thus the median of the values 13, 12, 10, 9, 15, 12, 8 is found after ranking the values, as follows:

Number	*Rank*	
15	1	
13	2	
12	3	
12	4	←Median = 12
10	5	
9	6	
8	7	

(Notice that the ranking used to find the median is slightly different from that used at other times. When a tie occurs, as with the two 12s in the above example, they are ranked one above the other rather than equal.)

If there is an even number of values, their median is the average of those on either side of the middle. Thus 6, 5, 4, 3, 2, 1, have median 3.5.

method of least squares, see LEAST SQUARES, METHOD OF.

method of semi-averages, see SEMI-AVERAGES, METHOD OF.

midrange, *n.* the ARITHMETIC MEAN of the smallest and largest values in a distribution. Thus, if x_1, x_2,. . .x_n are a set of numbers arranged in order of magnitude, then the midrange is

$$\frac{x_1 + x_n}{2}$$

It is half way between the largest and the smallest of the numbers.

Example: the midrange for the values 9, 9, 7, 7, 6, 5, 3, 3, 3, 2, is:

$$\frac{9+2}{2} = 5.5$$

The midrange of a sample is sometimes used as an ESTIMATOR for the population mean. If the parent distribution is uniform, midrange is a more EFFICIENT ESTIMATOR than the sample mean or the sample median, and is unbiased. See BIASED ESTIMATOR.

mode, *n.* the value of a variable which occurs most frequently. Example: a naturalist counted the eggs in each of 50 blackbirds' nests. The results were as follows:

Number of eggs	2	3	4	5	6
Frequency	1	10	26	11	2

The mode number of eggs is 4, which occurred 26 times; the next largest, 5, had a frequency of only 11.

If there are two modes, the distribution is said to be BIMODAL, three modes *trimodal*, and so on. The term, bimodal, is also applied to distributions with two distinct peaks even if they are not exactly equal in frequency.

If the data are grouped, the modal class is the class with the highest value of:

$$\frac{\text{Frequency}}{\text{Length of class interval}}$$

If the classes all have the same length of class interval the modal class is that with the most observations, the highest frequency. If, however, the intervals are of different lengths, this need not be the case. If a HISTOGRAM is drawn, the modal class will, however, always be that with the highest rectangle.

Example: the frequency distribution for the number of hours

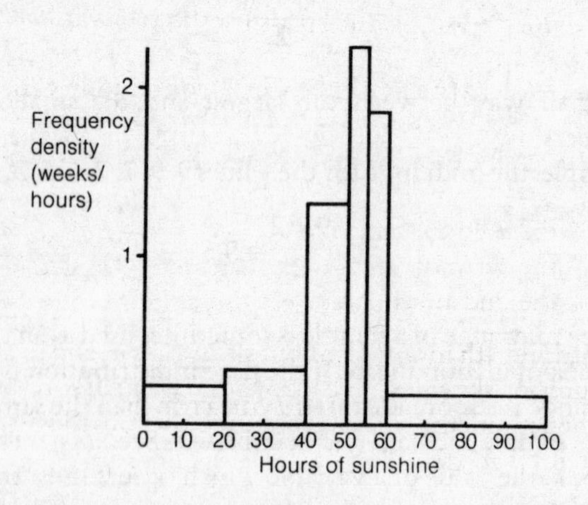

Fig. 67. **Mode.** Frequency distribution of hours of sunshine per week for a period of one year.

of sunshine per week over one year in a particular place was as follows:

<20	$20–40$	$40–50$	$50–55$	$55–60$	$60–100$
5	7	14	12	10	4

The modal class is not 40–55, even though this has the highest frequency. The class 50–55 has a higher frequency density as shown on the histogram (Fig. 67), and so this is the modal class.

model, *n.* a set of theoretical relationships which are intended to represent a real-life process, situation or problem. Many mathematical models involve considerable simplification and so may need to be refined in the light of experimental evidence.

modulus, see ABSOLUTE VALUE.

moment, *n.* the mean of a specified power of the deviations of all the values of a random variable in its frequency distribution. The power of the deviations indicates the order of the moment, and the deviations may be from the origin (giving a moment about the origin) or from the mean (giving a moment about the mean).

The nth moment of a distribution about its MEAN $\bar{x}$, (symbol: μ_n) is given by

$$\mu_n = \sum_i (x_i - \bar{x})^n \, p(x_i) \quad \text{(for a discrete variable)}$$

$$\mu_n = \int_{\text{all } x} (x - \bar{x})^n \, f(x)dx \quad \text{(for a continuous variable)}$$

It can easily be seen that, about the mean;

the 0th moment, $\mu_0 = 1$
the 1st moment, $\mu_1 = 0$ (since $\bar{x}$ is the mean)
the 2nd moment, $\mu_2 =$ the VARIANCE.

The 3rd and 4th moments about the means μ_3 and μ_4 are used in measures of skewness (see SKEW) and KURTOSIS respectively.

When moments are taken about the origin, the nth moment, denoted by μ_n', is given by

$$\sum_i x_i^n p(x_i) \qquad \text{or} \qquad \int_{\text{all } x} x^n \, f(x)dx$$

In this case:

the 0th moment, $\mu_0' = 1$

the 1st moment, $\mu_1' = Mean,\ \bar{x}$

the 2nd moment, $\mu_2' = \text{Variance} + \bar{x}^2$

moment-generating function, *n.* the moment-generating function of the distribution of the random variable X is given by:

$$E(e^{xt}) = \Sigma_i e^{x_i t} p(x_i) \text{ (for a DISCRETE VARIABLE)}$$

$$= \int_{\text{all } x} e^{xt} f(x) dx \text{ (for a CONTINUOUS VARIABLE).}$$

where $p(x_i)$ is the PROBABILITY DISTRIBUTION and $f(x)$ the PROBABILITY DENSITY FUNCTION. Symbol: $M_X(t)$.

When these expressions are expanded using

$$e^{tX} = 1 + tX + \frac{t^2X^2}{2!} + \frac{t^3X^3}{3!} + \ldots$$

the results can be separated into

$$\mu_0' + \mu_1' t + \mu_2' \frac{t^2}{e!} + \ldots + \mu'_n \frac{t^n}{n!} + \ldots$$

where μ'_n is the nth moment of the distribution about the origin.

If A and B are independent random variables with moment-generating functions $M_A(t)$ and $M_B(t)$, then the moment-generating function for the variable $(A+B)$ is given by:

$$M_{A+B}(t) = M_A(t).M_B(t)$$

The probability distributions (or probability density functions) of two random variables are the same if, and only if, they have the same moment-generating function. Thus the moment-generating function completely specifies a distribution. Not all distributions however possess moment-generating functions and, for this reason, the *characteristic function*, $E(e^{itx})$ is defined. This exists uniquely for every probability distribution.

Monte Carlo method, *n.* a method of finding the PROBABILITY DISTRIBUTION of the possible outcomes of a process or experiment by simulation. (Named after the casino at Monte Carlo, where systems for winning at games of chance such as roulette, etc., are often tried.)

There are processes which are too complicated to allow a theoretical analysis. In such cases, the probabilities of the various outcomes may be estimated by carrying out repeated simulations.

Example: In bridge, a player estimates the strength of a hand of 13 cards by counting 4 points for each ace, 3 for a king, 2 for a queen, 1 for a jack, and 0 for anything else. There are thus 40 points in a pack of cards (4 aces, 4 kings, 4 queens and 4 jacks) which are shared among the four hands dealt. The distribution of points is discrete, varying from 0 to the maximum possible for 13 cards, 37 (for 4 aces, 4 kings, 4 queens, and 1 jack), with mean 10. To say more about the distribution requires either a theoretical analysis, which would be very complicated, or an experiment, the Monte Carlo method. One possible experiment in this case would be to deal large numbers of hands from perfectly shuffled cards; another would be to carry out many runs of a computer simulation.

moving average, *n.* a form of AVERAGE which has been adjusted to allow for SEASONAL COMPONENTS or CYCLIC COMPONENTS. When a variable, like the cost of a sack of potatoes, or the number of unemployed, is graphed against time, there are likely to be considerable seasonal or cyclic components in the variation. These may make it difficult to see the underlying TREND. These components can be eliminated by taking a suitable moving average.

Example: the cost of a kilogram of dodo meat (averaged over each quarter) over the years 1980–1982 was as follows:

	1980	*1981*	*1982*
1st quarter	£2.00	£2.20	£2.60
2nd quarter	£3.00	£3.20	£4.00
3rd quarter	£2.00	£3.00	£3.40
4th quarter	£1.40	£2.20	£3.00

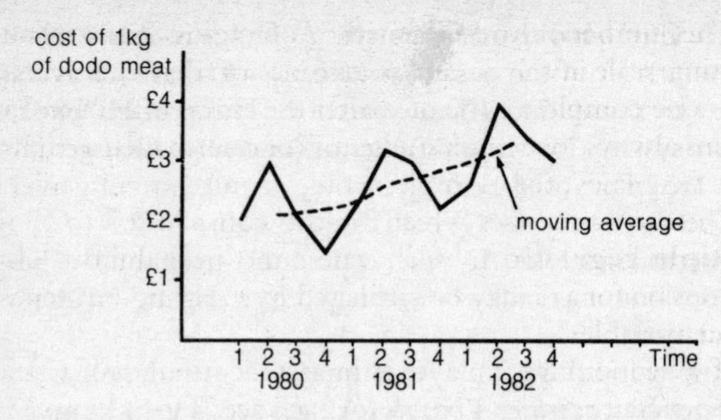

Fig. 68. **Moving average.** The cost of dodo meat plotted for each quarter year, and the moving average.

The seasonal component, which is very noticeable on the graph (Fig. 68), is eliminated when the *four-part moving average* is taken. The calculation for this is set out below, and the resulting average plotted on the graph (Fig. 68).

Start	'80,1	'80,2	'80,3	'80,4	'81,1	'81,2	'81,3	'81,4	'82,1
End	'80,4	'81,1	'81,2	'81,3	'81,4	'82,1	'82,2	'82,3	'82,4
Mean time	'80,2½	'80,3½	'81,½	'81,1½	'81,2½	'81,3½	'82,½	'82,1½	'82,2½
Price	2.00	3.00	2.00	1.40	2.20	3.20	3.00	2.20	2.60
	3.00	2.00	1.40	2.20	3.20	3.00	2.20	2.60	4.00
	2.00	1.40	2.20	3.20	3.00	2.20	2.60	4.00	3.40
	1.40	2.20	3.20	3.00	2.20	2.60	4.00	3.40	3.00
Total	8.40	8.60	8.80	9.80	10.60	11.00	11.80	12.20	13.00
Mean	2.10	2.15	2.20	2.45	2.65	2.75	2.95	3.05	3.25

This shows a clear upward trend in the cost of dodo meat.

This four-term moving average can be improved upon by taking the averages of successive values, a two-term moving average on the terms of the four-term moving average. This gives a weighted moving average over five quarters, $\frac{1}{8}$ (1, 2, 2, 2, 1), since:

$$\frac{1}{2}\left[\frac{1}{4}(a_0+a_1+a_2+a_3)+\frac{1}{4}(a_1+a_2+a_3+a_4)\right]$$
$$=\frac{1}{8}(a_0+2a_1+2a_2+2a_3+a_4)$$

where a_0, a_1, a_2, a_3 and a_4 are five successive quarterly values.

The number of parts in a moving average is clearly related to the time scale involved, and should be such that each average is over one complete cycle of variation. This is often, but by no means always, one year; studies of long-term changes in sunspot frequency, for example, would require moving averages to be taken over an 11-year cycle.

multiple regression, *n.* a technique for finding a best fit expression for a random variable which is dependent on several other variables.

If a random variable V is linearly dependent on several independent random variables $x_1, x_2, \ldots$, its value is given by:

$$V = \alpha_1 x_1 + \alpha_2 x_2 + \ldots + \epsilon$$

where α_1, α_2 are constants, and ϵ represents the residual variation. The values of $\alpha_1, \alpha_2, \ldots$ are estimated from the sample data, to minimize the sum of the squares of the values of the residual ϵ.

multiplication rule, *n.* the rule which states that, for two events A and B, the probability p of their both occurring is given by:

$$p(A \cap B) = p(A \mid B).p(B)$$
$$\text{or } p(A \cap B) = p(B \mid A).p(A)$$

If the events A and B are independent, this can be simplified to

$$p(A \cap B) = p(A).p(B)$$

Example: a man goes to the races and backs a horse, entirely at random, on the 2.00 and 2.45 races. The first race has 5 runners, the second 7. What is the probability that he picks both winners?

The first event, 'He picks a winner on the 2.00 race', has probability 1/5; the second event, 'He picks a winner on the 2.45 race', 1/7. The two events are independent.

The probability that he picks both winners is

$$\tfrac{1}{5} \times \tfrac{1}{7} = \tfrac{1}{35}$$

multivariate distribution, *n.* a distribution of more than one variable. If two variables are involved it is called bivariate, if three, trivariate and so on. Compare BIVARIATE DISTRIBUTION.

Consider a plantation of Sitka spruce trees. The joint distribution of height, base diameter and age for the trees is multivariate.

mutually exclusive, see EXCLUSIVE.

N

natural logarithm, see LOGARITHM.

negative binomial distribution, see PASCAL'S DISTRIBUTION.

nominal scale, see CATEGORICAL SCALE.

non-linear interpolation, see INTERPOLATION.

non-parametric tests of significance or **distribution-free tests of significance,** *n.* tests of significance which make no assumptions concerning the parent distribution. All tests involving the ranks of data are non-parametric. Examples are Kendall's rank correlation and Spearman's rank correlation (see CORRELATION COEFFICIENT), KRUSKAL-WALLIS ONE-WAY ANALYSIS OF VARIANCE, FRIEDMAN'S TWO-WAY ANALYSIS OF VARIANCE BY RANK, MANN WHITNEY U-TEST, the SIGN TEST, and WILCOXON MATCHED-PAIRS SIGNED-RANK TEST. Compare PARAMETRIC TESTS OF SIGNIFICANCE.

norm, *n.* an average level of achievement, performance or behaviour.

normal curve, Gaussian curve or **error curve,** *n.* the graph of the probability density function of the NORMAL DISTRIBUTION. It is a bell-shaped curve (Fig. 69), symmetrical about its mode (or mean, or median). Symbol: $\varphi(x)$

normal distribution, *n.* a continuous DISTRIBUTION of a random variable with its mean, median and mode equal. Symbol: $N(\mu, \sigma^2)$.

The probability density function $\varphi(x)$ of the normal distribution, with mean μ and standard deviation σ, is given by:

$$\varphi(x) = \left(\frac{1}{\sqrt{2\pi}\,\sigma}\right)\exp\left\{-\frac{(x-\mu)^2}{2\sigma^2}\right\}$$

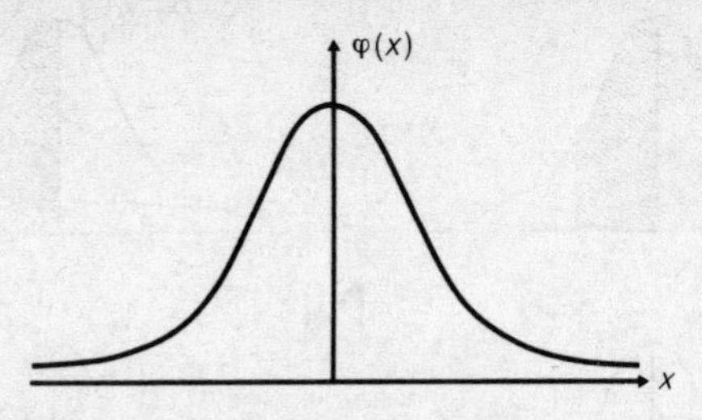

Fig. 69. **Normal curve.**

The distribution is denoted by $N(\mu, \sigma^2)$. When it is given in standardized form, with mean 0 and standard deviation 1, the distribution is denoted by $N(0, 1)$, and has probability density function given by:

$$\varphi(x) = \frac{1}{\sqrt{2\pi}} e^{-\frac{1}{2}x^2}$$

The graph of the probability density function of the normal distribution is a continuous bell-shaped curve, symmetrical about the mean, called the NORMAL CURVE or *error curve*.

The normal distribution is very important in statistics for two reasons:

(a) It is the distribution of many naturally-occurring variables, such as the heights of adult men in a town, the masses of carrots in a field, etc.

(b) The distribution of the means of samples drawn from most parent populations is normal or approximately so, when the samples are sufficiently large.

The area under the curve of the graph, representing the proportion of the population in question, is given in tables. The value of x to be used is the standardized variable,

$$\frac{x - \mu}{\sigma}$$

or the number of standard deviations from the mean. The graphs can be presented in various ways, as in Fig. 70. The tables on page 247 are of type (a), and the area in question in this case is denoted by $\Phi(x)$.

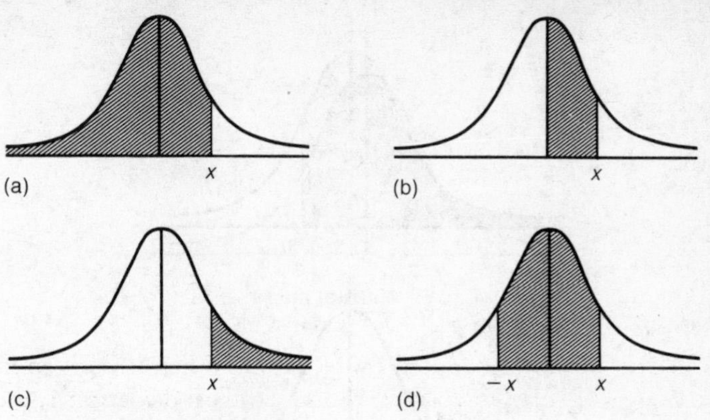

Fig. 70. **Normal distribution.** Graphs showing areas sometimes given in tables.

Example 1: in a certain town the mean height of adult men is 1.75m, with standard deviation .05m. There are 5,000 men in the town. How many should be
(a) Less than 1.80m;
(b) Greater than 1.85m;
(c) Between 1.80m and 1.85m?

(a) The area required is found as

$$\Phi\left(\frac{1.80-1.75}{.05}\right)=\Phi(1)=.8413$$

It is represented by the shaded area in Fig. 71a (overleaf). Since the probability of a man being less than 1.80m tall is .8413, the expected number of men in this category is:

$$.8413\times 5{,}000=4{,}206$$

(b) For men taller than 1.85m, the area required is that shaded in Fig. 71b, namely

$$1-\Phi\left(\frac{1.85-1.75}{.05}\right)=.0228$$

So the expected number of men over 1.85m tall is:

$$.0228\times 5{,}000=114$$

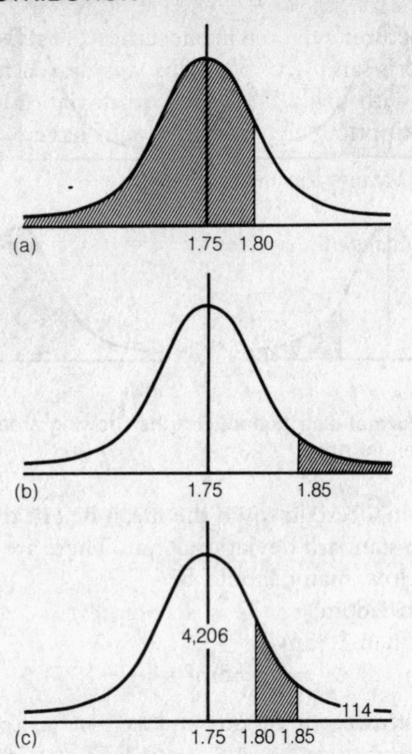

Fig. 71. **Normal distribution.** (a) Area $<$ 1.80 (b) Area $>$ 1.85 (c) 1.80 $<$ Area $<$ 1.85.

(c) The number of men between 1.80m and 1.85m in height (see Fig. 71c) is expected to be:

$$5{,}000 - (4{,}206 + 114) = 680$$

Example 2: a light-bulb maker claims that his bulbs are designed to have a mean life of 800 hours with standard deviation 40 hours. A customer finds that 50 bulbs that he bought had mean life 790 hours, and claims that they were substandard. The manufacturer replies that there is nothing wrong with his bulbs, but that the customer picked a bad sample. What is the probability that the manufacturer's claim is true?

The manufacturer's claim is in effect the NULL HYPOTHESIS: That the customer's sample of 50 bulbs was drawn from a parent population with mean 800, standard deviation 40 hours. On this assumption, the sample should have:

$$\textit{Mean} = \text{Parent mean} = 800$$

$$\textit{Standard deviation} = \frac{\sigma}{\sqrt{n}} = \frac{40}{\sqrt{50}} = 5.657$$

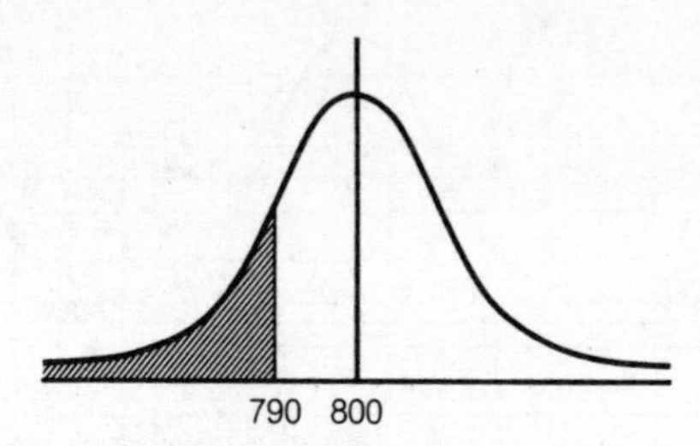

Fig. 72. **Normal distribution.** Area < 790

The area shaded in Fig. 72 is that required.

$$\Phi\left(\frac{790-800}{5.657}\right) = \Phi(-1.77)$$

Using the symmetry of the normal curve, the area under review is thus seen to be $1-\Phi(1.77) = .0384$. Thus, the probability of a sample having a mean as low as, or lower than, that recorded is .0384. This is less than 5%, and so gives reasonable grounds for being suspicious of the manufacturer's claim.

A useful way of checking if data fits a normal distribution is to plot the points on *normal probability graph paper*.

Example: 100 specimens of the common shrew were collected and weighed. Their masses were found to be as follows:

Mass (g)	5.0–5.5	5.5–6.0	6.0–6.5	6.5–7.0	7.0–7.5	7.5–8.0	8.0–8.5	8.5–9.0
Frequency (shrews)	1	10	24	31	18	10	4	2

The first step is to draw up a CUMULATIVE FREQUENCY table, and then to convert the cumulative frequencies into proportions of the total.

Mass	<5.0	<5.5	<6.0	<6.5	<7.0	<7.5	<8.0	<8.5	<9.0
Cum. fr.	0	1	11	35	66	84	94	98	100
Proportion	0	.01	.11	.35	.66	.84	.94	.98	1.00

These points are then plotted on normal probability graph paper (excepting the two end points, 0 and 1, which cannot be plotted); see Fig. 73. If the result is a straight line, the

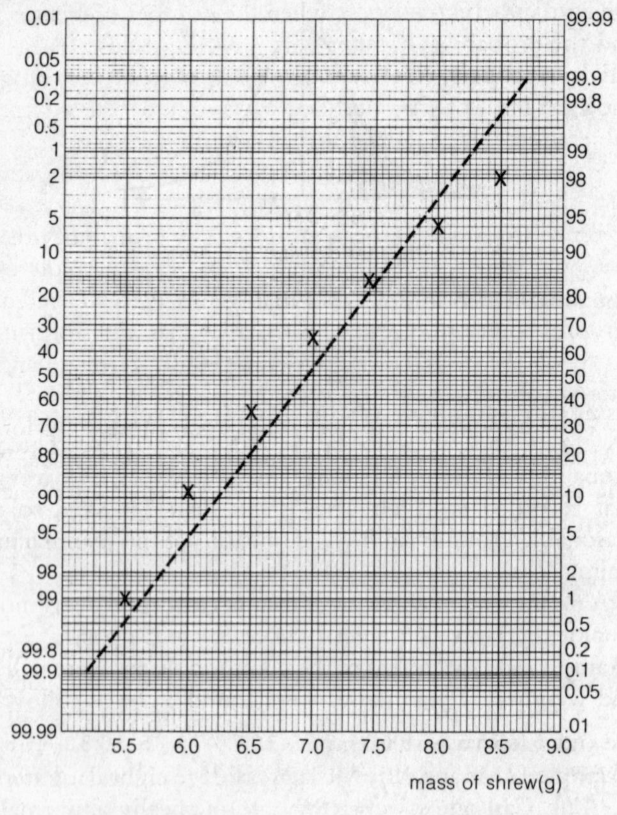

Fig. 73. **Normal distribution.** An example of data plotted on normal probability graph paper. The dashed line represents a normal distribution for the data; here, the data is positively skew, though with near-normal distribution.

distribution is normal. In this case it is not quite normal, but not very far from it either.

The shape of the curve of a non-normal distribution gives an indication of its SKEW. The distribution of the shrews is clearly positively skew (see Fig. 104).

A numerical test of how well data fit the normal (or any other) distribution can be carried out using the CHI-SQUARED TEST, χ^2 test.

The normal distribution may be used as an approximation to the BINOMIAL DISTRIBUTION when the number of trials is large and the probability neither very small nor close to 1.

null hypothesis, *n.* a hypothesis that is to be tested against another, but is to be nullified in favour of the alternative, subject to a given level of error. Symbol: H_0.

Statistical hypothesis testing is carried out by setting up a null hypothesis, H_0, and an *alternative hypothesis*, H_1. Data are then obtained from one or more samples. The probability that the data, or more extreme figures, are a chance result of the sampling, with the null hypothesis true, is then worked out. If this probability turns out to be smaller than the SIGNIFICANCE LEVEL of the test, the null hypothesis is rejected and the alternative hypothesis is accepted.

A null hypothesis is usually based upon the assumption that nothing special has occurred, no change has taken place. When one sample is involved the null hypothesis is often that the sample is drawn from a given parent population; for more than one sample, that the samples are drawn from the same parent population as each other. The alternative hypothesis is that something special has occurred, or a change has taken place.

There are two possible alternative hypotheses; that a change has taken place (leading to a two-tail test), and that a change has taken place in a particular direction (leading to a one-tail test); see ONE-AND TWO-TAIL TESTS.

Example 1: Mr and Mrs McTaggart have eight daughters and no sons. Can one say that they are medically abnormal?

Null hypothesis, H_0: There is nothing special about them. They are equally likely to have boys or girls. It is just by chance that their children are all girls.

Alternative hypothesis $H_1(1)$: They are more likely to have one sex of child than the other.
Alternative hypothesis $H_1(2)$: They are more likely to have girls than boys.
This can be expressed mathematically as follows. If p is the probability that a child is a girl, then

$$H_0: p=\tfrac{1}{2} \qquad H_1(1): p \neq \tfrac{1}{2} \qquad H_1(2): p>\tfrac{1}{2}$$

When the first alternative hypothesis is tested against the null hypothesis, the required probability is that of 8 boys or 8 girls out of 8 children, a two-tail test. This is thus:

$$2 \times (\tfrac{1}{2})^8 = \tfrac{1}{128}$$

For the second alternative hypothesis, the required probability is that of 8 girls;

$$(\tfrac{1}{2})^8 = \tfrac{1}{256}$$

These probabilities must then be interpreted in terms of a specified SIGNIFICANCE LEVEL.
Example 2: in 1933, the mean height of a sample of 40 adult men from a town was 1.73m. In 1983, a sample of 50 adult men from the same town had mean height of 1.78m. Can it be concluded that the men's height has changed?
Null hypothesis H_0: The mean height of adult men in the town has not changed. The observed differences are a consequence of the sampling.
Alternative hypothesis $H_1(1)$: The mean height of the men in the town has changed (a two-tail test).
Alternative hypothesis $H_1(2)$: The mean height of the men in the town has increased (a one-tail test).
If μ_1 is the 1933 parent mean, and μ_2 is the 1983 parent mean:

$$H_0: \mu_1=\mu_2 \qquad H_1(1): \mu_1 \neq \mu_2 \qquad H_1(2): \mu_2>\mu_1$$

To go further with this problem, it would be necessary to know further information concerning the standard deviation of the heights.

O

odds, *n.* the ratio of the probability of an event occurring to that of its not occurring.
Example: if the odds against Bolton Wanderers winning the FA Cup are 1:20, this means:

Probability (Bolton win): Probability (Bolton do not win)

$$= 1:20$$

and so

$$\text{Probability (Bolton win)} = \tfrac{1}{21}$$

$$\text{Probability (Bolton do not win)} = \tfrac{20}{21}$$

ogive, *n.* a CUMULATIVE FREQUENCY graph.

one- and two-tail tests, *n.* tests of a NULL HYPOTHESIS against different alternative hypotheses.

Statistical hypothesis testing is carried out by setting up a null hypothesis and testing it against an alternative hypothesis, of which there are two types:
(a) That the null hypothesis is false;
(b) That the null hypothesis is false in a particular direction.
The first leads to a *two-tail test*, the second to a *one-tail test.*
Example 1: A coin is tossed seven times, coming heads all seven, so that its fairness is called into doubt. Test, at the 1% SIGNIFICANCE LEVEL, (a) if it is biased; (b) if it is biased towards heads.
In both cases the null hypothesis is that the coin is unbiased. The probability of a head at any toss is $\frac{1}{2}$.
For (a), the alternative hypothesis is that the probability of heads is not $\frac{1}{2}$, leading to a two-tail test.

For (b), the alternative hypothesis is that the probability of heads is greater than $\frac{1}{2}$, leading to a one-tail test; see Fig. 74.

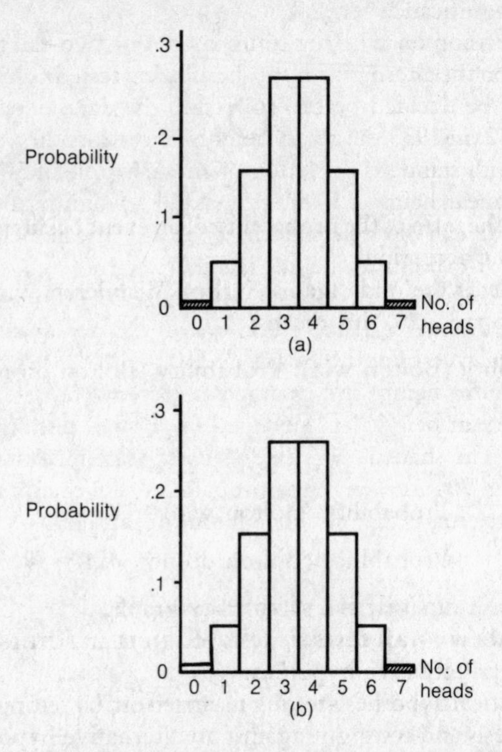

Fig. 74. **One- and two-tail tests.** (a) Two-tail test. (b) One-tail test.

For (a), the probability of a result as extreme as 7 heads is the probability of 7 heads or 7 tails:

$$(\tfrac{1}{2})^7 + (\tfrac{1}{2})^7 = .0156$$

Since this is greater than 1%, there is no reason to reject the null hypothesis at this level. It cannot be claimed that the coin is biased at the 1% significance level.

For (b), the probability of 7 heads is:

$$(\tfrac{1}{2})^7 = .0078$$

Since this is less than 1%, the null hypothesis is rejected at this level. It can be claimed that the coin is biased towards heads at the 1% significance level.

The decision on whether to use a one- or two-tail test thus depends on the alternative hypothesis being tested. This would normally be decided before collecting the data.

Example 2: in 1933 the mean heights of the men in a city was 1.75m, with standard deviation .07m. A sample of 50 men in 1983 had mean height 1.77m. Test, at the 5% significance level, whether it can be claimed that: (a) the mean height has changed; (b) the mean height has increased.

In comparing the 1983 sample with the 1933 population, the null hypothesis is that the mean height has not changed.

The alternative hypotheses are:

(a) The mean height has changed (a two-tail test);

(b) The mean height has increased (a one-tail test).

Case (a). The shaded areas (see NORMAL DISTRIBUTION) of the graph (Fig. 75a) represent the probability of a chance result at least as extreme as that obtained from the sample.

$$\mathrm{SD} = \frac{\sigma}{\sqrt{n}} = \frac{.07}{\sqrt{40}} = .011$$

$$\text{Shaded area} = 2\left[1 - \Phi\left(\frac{1.77 - 1.75}{.011}\right)\right] = .07$$

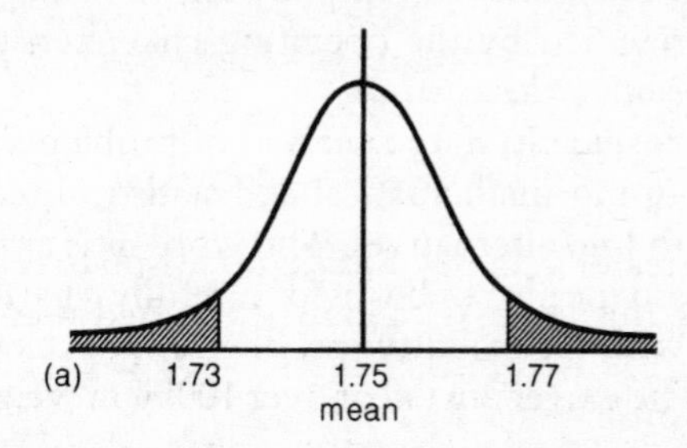

Fig. 75(a). **One- and two-tail tests.** Case (a), a two-tail test.

Since .07 > 5%, there is no reason to reject the null hypothesis at the 5% level. The mean height of men has not changed.

Case (b). In this case the shaded area (Fig. 75b) is:

$$\text{Shaded area} = 1 - \Phi\left(\frac{1.77 - 1.75}{.011}\right) = .035$$

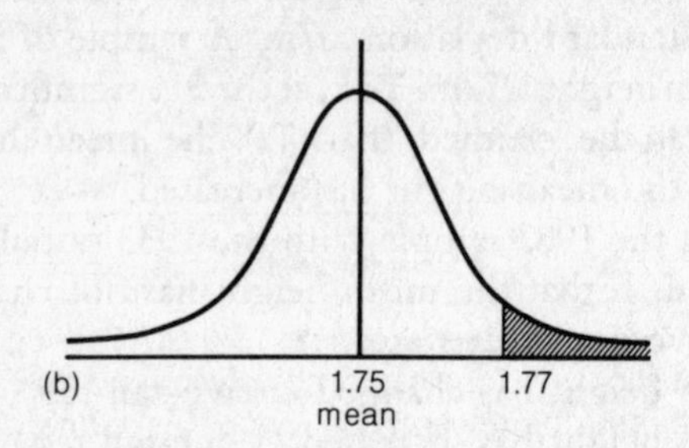

Fig. 75(b). **One- and two-tail tests.** Case (b), a one-tail test.

Since .035 < 5%, the null hypothesis is rejected at the 5% significance level. The alternative hypothesis, that the mean height has increased, is accepted.

one-way analysis of variance, see ANALYSIS OF VARIANCE, KRUSKAL–WALLIS ONE-WAY ANALYSIS OF VARIANCE.

operating characteristic curve, *n.* in quality control, the graph of the probability of accepting a batch from which a sample has been drawn for inspection, against the true proportion of defectives for the batch (Fig. 76). This curve is dependent on the particular sampling scheme being used. The function represented by the operating characteristic curve is called the *operating characteristic.*

operational research, *n.* the analysis of problems in business and industry using mathematical and statistical techniques.

or, *conj.* used to join alternatives. The word *or* is ambiguous in English and so needs to be used carefully, particularly in connection with probability problems. For example, the condition, 'If he is over 2m tall or over 100kg in weight. . .' can mean:

(a) If he is over 2m tall, or over 100kg in weight, or both; in

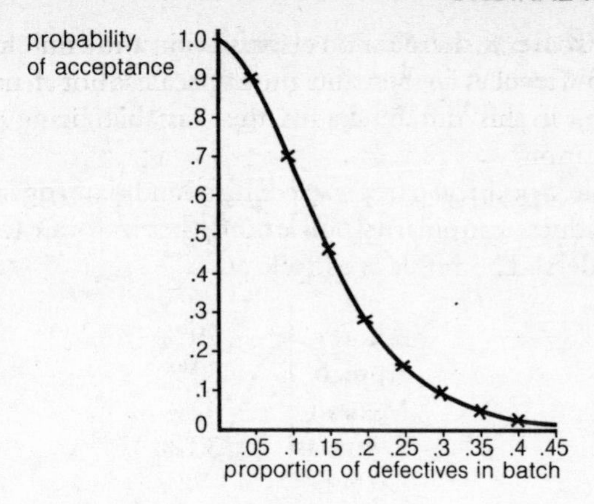

Fig. 76. **Operating characteristic curve.** The operating characteristic curve for a sampling scheme in which sample size is 12, and the acceptance number is 1 (the batch is accepted if 0 or 1 are faulty).

this case the original *or* is inclusive because the possibility of both events occurring is included;

(b) If he is either over 2m tall, or over 100kg in weight, but not both. The *or* is now exclusive, because the case when both events occur is excluded.

For events A and B, these two meanings can be illustrated on VENN DIAGRAMS, as in Fig. 77.

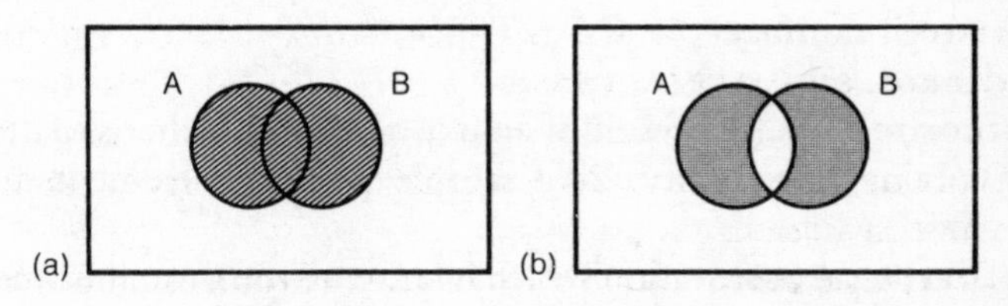

Fig. 77. **Or.** (a) Venn diagram to illustrate the inclusive case of *or*, $A \cup B$. (b) Venn diagram to illustrate the exclusive case of *or*, $(A \cup B) \cap (A \cap B)'$.

ordinal number, *n.* a number denoting order, quality or degree in a group, such as first, second, third. Compare CARDINAL NUMBER.

ordinal scale, *n.* a scale with classification and rank. However, equal differences in position on the scale do not necessarily represent equal differences in the variable being used for classification.

Example: a political party is selecting a candidate for an election and conducts a popularity poll among the electorate for the five contenders. The result is as follows:

Allotey	29%
Appiah	32%
Mensah	2%
Twumasi	30%
Zwennes	7%

They are then set in order:

1	Appiah
2	Twumasi
3	Allotey
4	Zwennes
5	Mensah

This 1-to-5 scale is an ordinal scale. Number 1 is more popular than number 2 and so on. However, the difference between numbers 1 and 3 (3%) is not necessarily the same as that between numbers 3 and 5 (27%).

ordinate, see ABSCISSA, GRAPH.

outcome, *n.* **1.** the result of an experiment or other situation involving uncertainty. **2.** a sample point, an element in a SAMPLE SPACE.

outlier, *n.* an observation which is far removed from the others in the set. An outlier may be an observation from a different parent population, or it may be the result of experimental error, or it may be a genuine result. Since outliers can have a considerable influence on test statistics, they should be examined carefully before being accepted.

The diagram (Fig. 78) represents a number of stars plotted according to their emission of visible light and their surface temperatures (logarithmic scale). Such a graph is called a *Hertzsprung-Russell diagram.*

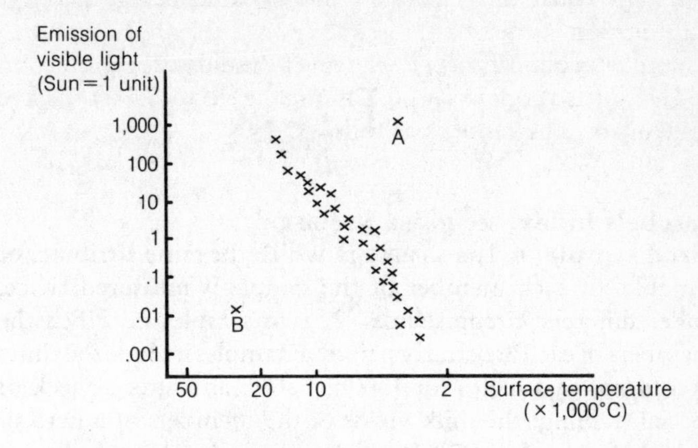

Fig. 78. **Outlier.** Hertzsprung-Russell diagram for a number of stars. A and B are outliers.

Clearly the stars A and B are outliers. The reason for this is that all the other stars are in the phase of their evolution when they are burning hydrogen into helium; they are main sequence stars. Star A, however, is in a later stage of its evolution, and is a red giant. Star B is in a still later stage and is a white dwarf. Thus, in this case, the two outliers belong to essentially different parent populations. See EXPLORATORY DATA ANALYSIS.

P

Paasche's Index, see INDEX NUMBER.

paired sample, *n.* **1.** a SAMPLE in which the same attribute, or variable, of each member of the sample is measured twice, under different circumstances. **2.** two samples in which the members of each are clearly paired. Examples include the times of a group of athletes for 1500m before and after a week of special training; the milk yields of the members of a herd of cows before and after being fed a particular diet; intelligent quotient measurements on pairs of identical twins, and measurements of wear on left and right shoes.

The difference in the value of the variable for each matched pair is worked out; these figures then form a single sample for an appropriate test. This could be a test of the NULL HYPOTHESIS that the mean difference is zero, carried out using the NORMAL DISTRIBUTION or the paired sample T-TEST. The SIGN TEST is concerned with median difference. The WILCOXON MATCHED-PAIRS SIGNED-RANK TEST uses the ranks of the sizes of the differences, giving the ranks the signs of the differences.

paired-sample test, *n.* a test on the differences of the random variable measured in a paired sample, usually of the Null Hypothesis that the mean value of the difference is zero.

Example: ten hens, labelled A to J, had their egg yield measured over a four-week period. They were then fed Superlay pellets, and their yields recorded over the next four weeks. Could it be claimed, at the 5% SIGNIFICANCE LEVEL, that their yield had altered?

Null hypothesis: There is no change in their yield. The differences observed are a sample drawn from a parent population with mean 0.

Alternative hypothesis: There is a difference in their yield.
Test: Paired sample t-TEST, two-tail (see ONE- AND TWO-TAIL TESTS).
Critical value: $(10-1=9$ degrees of freedom, 5% significance level) 2.26.

		A	B	C	D	E	F	G	H	I	J
YIELD (eggs)	*Before*	26	24	15	23	14	18	17	28	18	19
	After	28	24	18	21	19	21	17	27	24	26
	Difference (after-before)	+2	0	+3	−2	+5	+3	0	−1	+6	+7

The *mean* of the observed differences is

$$\frac{2+0+3-2+5+3+0-1+6+7}{10}=2.3$$

Estimated parent standard deviation

$$=\sqrt{\left(\frac{(-.3)^2+(-2.3)^2+.7^2+(-4.3)^2+(2.7)^2+(.3)^2+(-3.3)^2+3.7^2+4.7^2}{10-1}\right)}$$

$=3.05$

Estimate of the standard deviation of the sampling distribution of the means of samples of size 10

$$\frac{\sigma}{\sqrt{n}}=\frac{3.05}{\sqrt{10}}=.964$$

The statistic t is then calculated as

$$t=\frac{2.3-0}{.96}=2.38$$

As $2.38>2.26$ (the critical value), the null hypothesis is rejected at the 5% level, and it is concluded that the hens' egg yield is altered by the new pellets. (Since the values 2.38 and 2.26 are close it might be wise to conduct further tests to confirm this conclusion.)

If the sample size had been large, the test would have been conducted using the normal distribution rather than the t-distribution. Other tests on paired samples include the SIGN TEST and WILCOXON MATCHED-PAIRS SIGNED-RANK TEST.

The term 'matched-sample' can be applied to more than two matched samples. FRIEDMAN'S TWO-WAY ANALYSIS OF VARIANCE BY RANK is used on data from several matched samples.

Care must be taken not to confuse a paired-sample test with a TWO-SAMPLE TEST. A two-sample test is used when the individual members of the two samples are not the same.

paradox, *n.* a seemingly absurd or self-contradictory statement.

parameter, *n.* a value, known or unknown, applied to a parent population rather than a sample, which is used to define a statistical model, usually a theoretical distribution. For example, a NORMAL DISTRIBUTION is usually defined by two parameters, its mean and variance. Thus the normal distribution with mean 20 and variance 7 may be written N(20,7).

A BINOMIAL DISTRIBUTION is also defined by two parameters, the number of TRIALS n and the probability of success in each one, p. It is sometimes written B(n,p). A POISSON DISTRIBUTION needs just one parameter, the mean μ. It is written $P(\mu)$.

The choice of parameters is not unique.

Example: An EXPONENTIAL DISTRIBUTION may be written as

$$\lambda e^{-\lambda x} \text{ or } \frac{1}{\theta} e^{-x/\theta}$$

with parameters λ or θ respectively. The mean of the distribution is θ, which is equal to $1/\lambda$. The sample mean is an unbiased estimator of θ, but there is no linear unbiased estimator of λ.

Compare STATISTIC.

parametric tests of significance, *n.* tests of significance which assume that the population distribution has a particular form (e.g., normal) and involve hypotheses about population parameters. The parameters most commonly used in such tests are mean, standard deviation and, in the case of a bivariate distribution, covariance.

Examples of parametric tests are those using NORMAL DISTRIBUTION, the t-TEST, product-moment correlation (see CORRELATION COEFFICIENT), and the F-TEST. Compare NON-PARAMETRIC TESTS OF SIGNIFICANCE.

parent population, *n.* the population from which a SAMPLE is drawn, any member of which could have been selected (Fig. 79); sometimes this is just referred to as the *population*. See also SAMPLE SPACE.

There are times when an underlying parent population may be assumed, even though it does not strictly exist. The recovery times of 20 people who have contracted a new disease form a sample from the (non-existent) parent population which would be formed if many other people caught the disease.

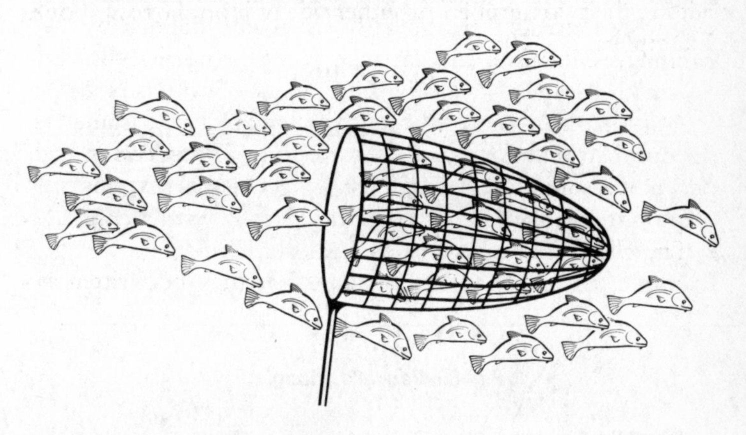

Fig. 79. **Parent population.** Showing the relationship between a parent population and a sample taken from it.

Pascal's distribution or **negative binomial distribution,** *n.* the distribution with PROBABILITY DISTRIBUTION given by:

$$p(x) = {}^{x-1}C_{x-n}\, p^n q^{x-n}$$

for $x > n$ where $0 < p < 1$ and $q = 1 - p$. This is the probability distribution for the number of BERNOULLI TRIALS x needed to achieve n successes. Each TRIAL has probability p of success, q of failure.

The mean and variance of this distribution are given by:

$$Mean = \frac{nq}{p}$$

$$Variance = \frac{nq}{p^2}$$

When $n = 1$, Pascal's distribution becomes the GEOMETRIC DISTRIBUTION.

Pascal's triangle, *n.* a triangular diagram consisting of rows of numbers (Fig. 80). It is named after Blaise Pascal (1623–62), French philosopher and mathematician. Each row is formed from the one above it by putting a 1 at each end, and then placing new numbers at regular intervals, each calculated by adding the two numbers on either side of it on the row above. Example: . . 5 10 . .
. . 15 . . $15 = 5 + 10$

```
            1
          1   1
        1   2   1
      1   3   3   1
    1   4   6   4   1
  1   5  10  10   5   1
1   6  15  20  15   6   1
•   •   •   •   •   •   •   •
```

Fig. 80. **Pascal's triangle.**

Pascal's triangle may be used for calculating BINOMIAL COEFFICIENTS. This relationship is illustrated by comparing row four of Pascal's triangle:

1 4 6 4 1

and the binomial expansion for $(x + y)^4$:

$$(x + y)^4 = 1x^4 + 4x^3y + 6x^2y^2 + 4xy^3 + 1y^4$$

Pearson's measure of skewness, see SKEW.

Pearson's product-moment correlation coefficient, see CORRELATION COEFFICIENT.

peer group, *n.* a social group composed of individuals of approximately the same age or status.

percentage, *n.* proportion or rate per hundred parts. Symbol: %. To write a number as a percentage, all that is needed is to multiply it by 100.

Example: write 42 out of 60 as a percentage.

$$\tfrac{42}{60} \times 100 = \tfrac{7}{10} \times 100 = 70\%$$

Thus 1 = 100%, 2 = 200%, 1/5 = 20%, and .003 = .3%.

percentage bar diagram, *n.* a method of showing a population which is divided into several sets. In this sort of diagram, a single bar represents the whole population (100%) and is divided into parts in the correct proportions (Fig. 81).

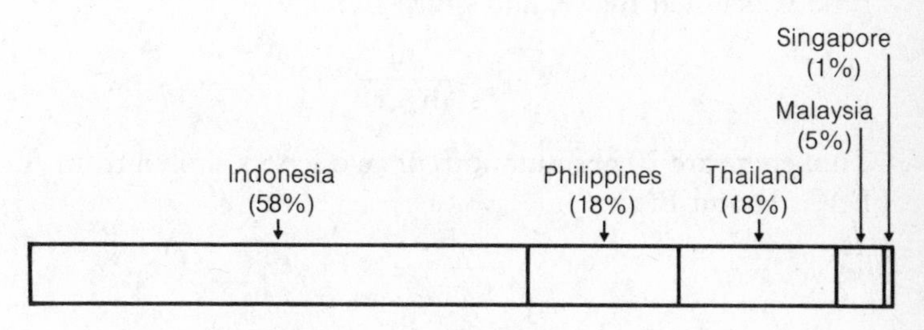

Fig. 81. **Percentage bar diagram.** Populations of the countries in the Association of Southeast Asian Nations (ASEAN). Source: *Philip's Certificate Atlas*, 1978.

percentage frequency distribution, *n.* a frequency distribution expressed as a percentage of the whole.

Example: daily rainfall in one year at a weather station.

Rainfall (cm)	$0 \leqslant\ < .25$	$.25 \leqslant\ < .5$	$.5 \leqslant\ < .75$	$.75 \leqslant\ < 1.0$
Frequency	157	66	52	37
% frequency (nearest 1%)	43	18	14	10

Rainfall (cm)	$1.0 \leqslant\ < 1.25$	$1.25 \leqslant\ < 1.5$	$1.5 \leqslant\ < 1.75$	$1.75 \leqslant\ < 2.0$
Frequency	28	18	4	3
% Frequency (nearest 1%)	8	5	1	1

percentile, *n.* one of 99 actual or notional values of a variable dividing its distribution into 100 groups with equal frequencies. See also QUANTILE.

permutation, *n.* an ordered arrangement of the numbers, terms, etc., of a set into specific groups. The number of permutations of n objects, all different, is n! Thus there are six permutations of the letters A, B, C, namely:

A B C	B A C	C A B
A C B	B C A	C B A

The number of permutations of n elements selected r at a time is denoted by $^{n}P_{r}$ and given by:

$$^{n}P_{r} = \frac{n!}{(n-r)!}$$

Thus there are 20 permutations of two letters chosen from A, B, C, D and E; i.e.,

$$\frac{5!}{(5-2)!}$$

A B	B A	C A	D A	E A
A C	B C	C B	D B	E B
A D	B D	C D	D C	E C
A E	B E	C E	D E	E D

The number of permutations of n objects, p of one type, q of another, r of another, etc., is given by

$$\frac{n!}{p!\ q!\ r!\ \ldots}$$

Thus the permutations of the letters POSSESSES number

$$\frac{9!}{5!\ 2!\ 1!\ 1!} = 1512$$

The term permutation is often misused by those who fill in the football pools and use it to mean combination.

pictogram, *n.* a diagram in which the frequency of an item is indicated by a number of identical pictures. Quantity should be shown in a pictogram by having more of the diagrams, not by enlarging them (Fig. 82); this can lead to confusion, as it is often unclear if the scale factor of the increase is that for lengths, areas or volumes.

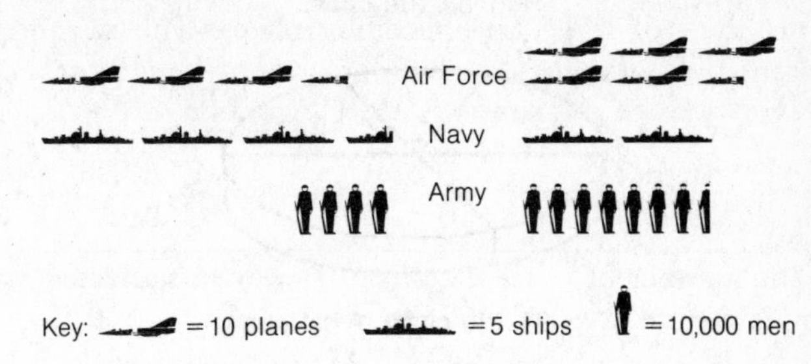

Fig. 82. **Pictogram.** The forces of two warring states.

pie chart, *n.* a circular diagram divided into sectors proportional to the frequencies or magnitudes of the items represented.

Example: a company spent £6 million on advertising one year, divided up as follows:

Television	£3m
Sports sponsorship	£1m
Newspaper adverts	£1.5m
Posters	£0.5m

In the pie chart produced (Fig. 83), the total of £6m = 360°, so £1m = 60°. Thus,

Television	:	$3 \times 60° = 180°$
Sports sponsorship	:	$1 \times 60° = 60°$
Newspaper adverts	:	$1.5 \times 60° = 90°$
Posters	:	$0.5 \times 60° = 30°$

When a pie chart is drawn as a disc, it is called a *slit chart*, Fig. 84.

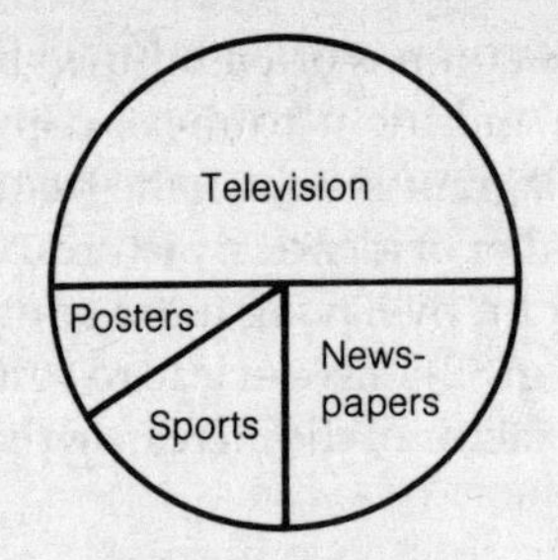

Fig. 83. **Pie chart.**

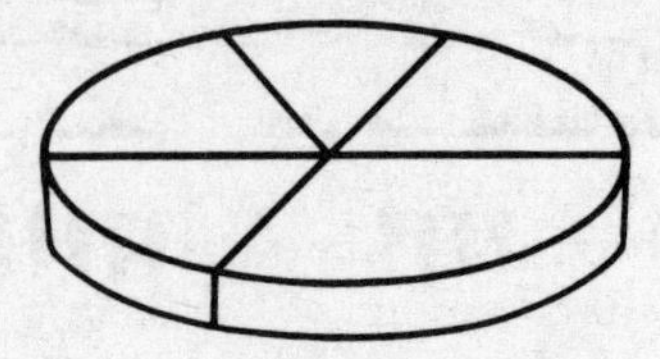

Fig. 84. **Pie chart.** A slit chart.

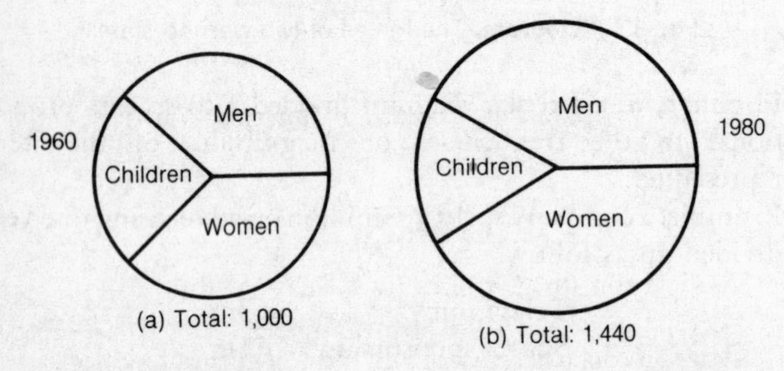

Fig. 85. **Pie chart.** Population of Little Puddlington, 1960 and 1980.

When two pie charts are drawn for comparison, the sizes of the two populations must be proportional to the areas and not the radii of the circles.

The two pie charts, A and B, shown in Fig. 85 have radii in the ratio 5:6, and so areas in the ratio $\pi \times 5^2, \pi \times 6^2 = 25:36$. Thus, if A represents a total of 1,000 people, B represents

$$\frac{36}{25} \times 1{,}000 = 1{,}440 \text{ people}$$

pilot survey, *n.* a survey carried out, usually on a small scale, before a larger survey. It is carried out to check the appropriateness of the intended procedures, particularly with regard to sampling and design of questionnaires. If little is known in advance, the term *exploratory survey* is sometimes used.

point estimation, *n.* estimation of a parameter of a parent population as a single value. This contrasts with *interval estimation*, where the parameter is estimated to lie within an interval. Thus, if the following numbers are sample values of a variable, 40, 52, 68, 70, 80, the parent mean could be estimated as 62, the sample mean. This would be a point estimate. It could also be estimated as lying, say between 42.3 and 81.7, the 95% CONFIDENCE INTERVAL; that would be an interval estimate.

point quadrat, see QUADRAT.

Poisson distribution, *n.* the DISTRIBUTION (symbol: $P(\mu)$) with probability distribution

$$p(r) = e^{-\mu}\frac{\mu^r}{r!}$$

with $r \in \{\text{integers} \geqslant 0\}$

Thus the probability-generating function for the Poisson distribution is:

$$\sum_r e^{-\mu}\frac{(\mu t)^r}{r!}$$

The Poisson distribution gives the probability of an event occurring 0, 1, 2, ..., r times under conditions when the probability p of its occurrence is small, but the number n of occasions on which it can occur is large. It is specified in terms of its mean, or expectation, μ, which is given by $\mu = np$. The Poisson distribution may be looked on in two ways:

(a) An approximation to the BINOMIAL DISTRIBUTION when n is large and p small.

(b) A distribution in its own right, which can be used when the mean is known but p and n need not be. In some situations it may be impossible to define p and n realistically, but their product $np = \mu$ can be well defined. If, for example, a football team scores 84 goals in 42 matches, $\mu = 84/42 = 2$ goals/match.

We cannot, however, say anything about either the probability of their scoring at a particular moment of a match or the number of such moments. In such a case the Poisson distribution is clearly of more significance than just an approximation to the binomial distribution.

The probability of exactly r occurrences is given by:

$$e^{-\mu}\frac{\mu^r}{r!}$$

The variance of the Poisson distribution is equal to its mean, μ, or np.

The Poisson distributions for $\mu = 1, 2$ and 3 are shown in the graphs of Fig. 86.

Typical situations which can be modelled by the Poisson distribution are the numbers of goals scored by a football team in a match; the number of air accidents occurring on any day; the number of telephone calls coming into an exchange in a period of one minute.

Example: in a large maternity hospital there are on average 2 pairs of twins every week. Find the probabilities of 0, 1, 2, 3 sets of twins being born in any week.

The mean incidence of twins/week, $\mu = 2$.

No. of pairs of twins	*Probability*
0	$e^{-2} \times 1 = .135$
1	$e^{-2} \times 2 = .271$
2	$e^{-2} \times \frac{2^2}{2!} = .271$
3	$e^{-2} \times \frac{2^3}{3!} = .180$

Calculations using the Poisson distribution can be rather cumbersome if the probability required is that of 'less than', 'more than', 'at most', or 'at least *X* occurrences' where *X* is not particularly small. For this reason the *Poisson probability chart* is published (Fig. 87. Compare GRAPH). It allows one to find the probability p of at least c occurrences for any mean. Thus, in the example of the twins, the question could have been asked,

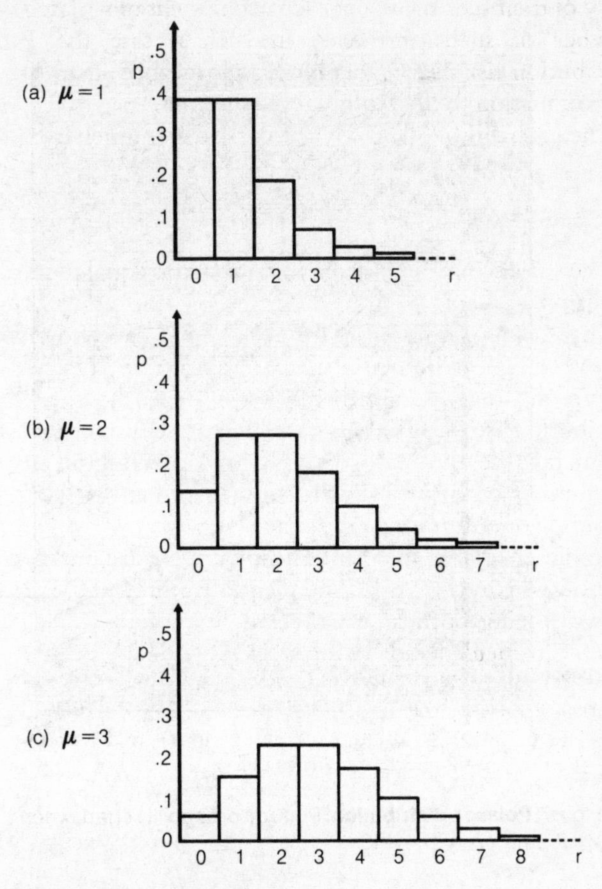

Fig. 86. **Poisson distribution.** (a) $\mu=1$. (b) $\mu=2$. (c) $\mu=3$.

'What is the probability of at least 4 pairs of twins in any week?'

The answer from the previous work is

$$1-(.135+.271+.271+.180)=.143$$

This can be found on the chart by finding the value of p where the line $\mu=2$ and the curve $c=4$ cross.

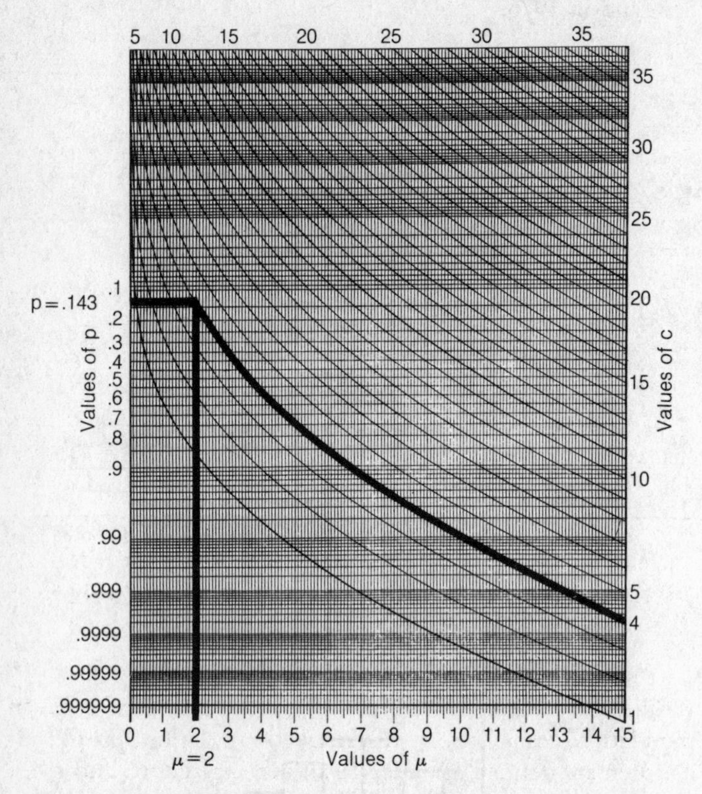

Fig. 87. **Poisson distribution.** Poisson probability chart; when $\mu = 2$, and $c = 4$, $p = .143$.

polar graph, see GRAPH.

population, *n.* **1.** the entire aggregate of individuals or items from which SAMPLES are drawn; any collection of things, people or numbers which is of interest. **2.** PARENT POPULATION.

population profile, *n.* a type of BAR CHART used to display the population age structure of a country (or city, region, etc.). The population figures are divided into age groups, and into male and female. The scale for the bars may be for the actual numbers or for their percentages. The chart in Fig. 88 shows that, for example, there were 180,000 boys aged 0 to 4 living in Scotland in 1976.

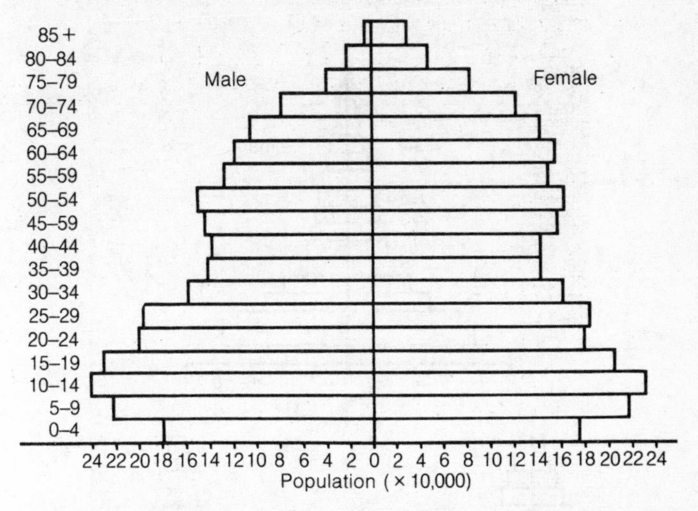

Fig. 88. **Population profile.** The population age structure of Scotland in 1976.

There are three main types of age structure of a given population; each has a distinctive population profile. If children are defined as under 16 and the aged as 65 and over, these can be categorized as follows (Fig. 89):
Progressive: over 45% children, and under 10% aged.
Regressive: under 30% children, and over 15% aged.
Stationary: 35–40% children, and 10% aged.
An *intermediate* structure is one changing from one type to another.

posterior probability, see PRIOR PROBABILITY.

power, *n.* the probability that a statistical test rejects the NULL HYPOTHESIS at a given value of the parameter being tested. If a null hypothesis is true and it is being tested at the 5% SIGNIFICANCE LEVEL, there is a 5% probability of rejecting it. Thus the power is .05 at the true value of the parameter being estimated. When a type 2 error occurs a false null hypothesis is accepted; so in that case

$$\text{Power} = 1 - \text{probability of a Type 2 error.}$$

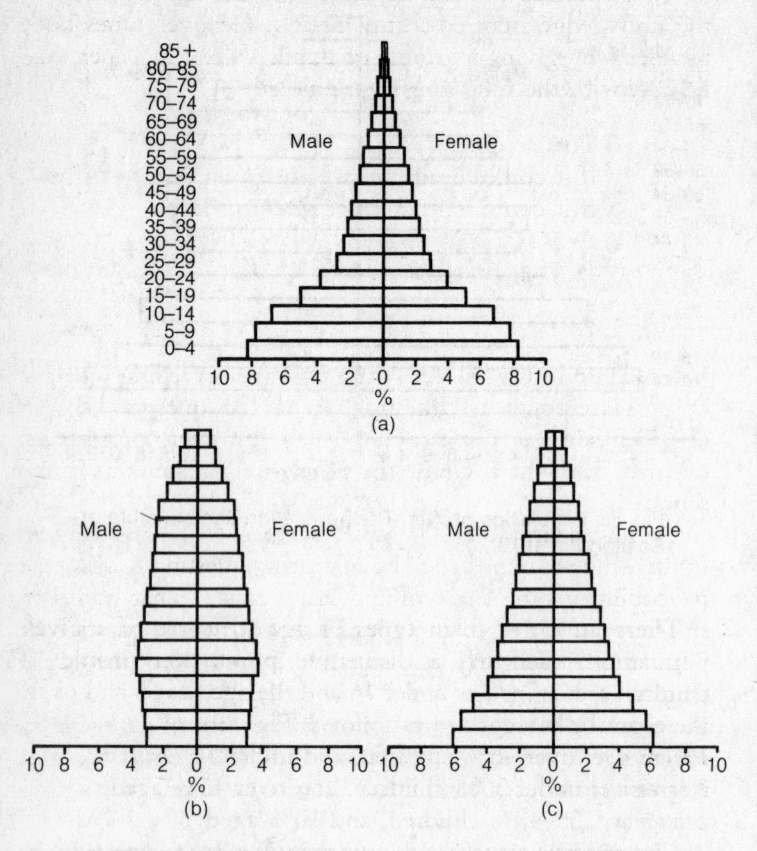

Fig. 89. **Population profile.** (a) Progressive. (b) Regressive. (c) Stationary.

principle of the irrelevant question, *n.* a technique for obtaining information about a population which might be damaging or incriminating to individuals.

Example: in a school smoking is disallowed. Anyone caught doing so is automatically expelled. Teachers who fail to report pupils they catch smoking are liable to instant dismissal. A teacher trying to find out what percentage of the pupils do nonetheless smoke faces two difficulties. The pupils are hardly likely to admit to him that they do, and if anyone should do so the knowledge may cost him his job. He overcomes both problems by giving a sample of pupils a sheet of paper, one Friday, with the following instructions:

> Toss a coin.
> If it comes heads, answer question A.
> If it comes tails, answer question B.
> A. Is it Friday today?
> B. Do you smoke?
>
> Yes/No

When a pupil answers 'Yes', there is no way of telling whether it is in response to question A or to question B. So confidentiality is not broken. It is, however, possible to estimate from the response the proportion of smokers in the school.

Suppose a sample of 100 pupils produces 65 Yes answers. 50 of these may be expected to be answering question A, since the probability of the coin coming heads is 1/2. That leaves 15 smokers out of the 50 who got tails, and answered question B. The teacher concludes that approximately 30% of the pupils smoke.

prior probability, *n.* (the prior probability of an event A, with respect to event B), the probability that event A occurs if it is not known if event B has occurred or not. This is thus p(A).

The *posterior probability*, for the same pair of events, is the probability that event A occurs if it is known that event B has occurred. This is written $p(A \mid B)$.

Example: two dice are thrown, first one then the other. Before either is thrown the probability of a total of at least 9, the prior probability, is 5/18. If, however, the first die is seen to land 6, the probability of a total of at least 9, the posterior probability, is 2/3.

In this example, we have

Event A: the total is 9
Event B: the first die is 6

The probability of total 9, 10, 11 or 12 p(A) is thus:

$$p(A) = \frac{4}{36} + \frac{3}{36} + \frac{2}{36} + \frac{1}{36} = \frac{10}{36} = \frac{5}{18}$$

The probability p(A | B) that the second die shows 3, 4, 5, or 6, is

$$p(A \mid B) = \frac{4}{6} = \frac{2}{3}$$

If the prior and posterior probabilities are equal, events A and B are independent.

probability, *n.* a measure of the relative frequency or likelihood of occurrence of an event. Symbol: p. Values are derived from a theoretical distribution or from observations. Probability is a number between 0 and 1. Probability 0 means impossibility, 1 is certainty.

For a discrete distribution, probability is defined as:

$$\frac{\text{the number of required outcomes}}{\text{the total number of possible outcomes}}$$

Thus the probability of drawing a spade from a pack of 52 playing cards is $\frac{13}{52} = \frac{1}{4}$, because
The number of required outcomes = 13 (spades)
The total number of possible outcomes = 52 (cards).

For a continuous variable, the probability is the relevant area under the graph of its probability density function.

In the case shown in Fig. 90, the probability density function f(x) is:

$$f(x) = \frac{3}{32}(4x - x^2)$$

the probability that x lies between 1 and 2 is the shaded area, which equals 11/32. This area represents

$$\frac{\text{the frequency of results between 1 and 2}}{\text{the total frequency of all results}}$$

See also IMPOSSIBLE, BAYESIAN CONTROVERSY.

probability density function, *n.* the function $f(x)$ of a continuous random variable X, such that

$$\int_a^b f(x)\,dx$$

is the probability that X lies between a and b.

In the diagram (Fig. 90) the probability that x lies between 1 and 2 is given by the shaded area, 11/32.

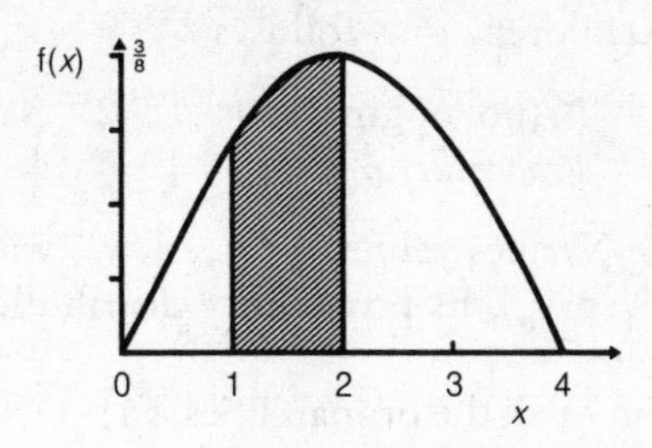

Fig. 90. **Probability density function.**

$$f(x) = \frac{3}{32}(4x - x^2)$$

If $f(x)$ is a probability density function, it must obey two conditions:

(a) that the total probability for all possible values of X is 1.

$$\int_{\text{all } x} f(x)\,dx = 1$$

(b) that the probability can never be negative:

$$f(x) \geqslant 0 \text{ for all } x$$

The mean (or expectation) and variance of x are given by:

$$\textit{Mean} = \int_{\text{all } x} x f(x)\,dx$$

$$\textit{Variance} = \int_{\text{all } x} x^2 f(x)\,dx - \text{mean}^2$$

At the mode of the distribution, $f'(x)=0$, where f' is df/dx.

The median of the distribution is the value of x such that

$$\int_{-\infty}^{x} f(x)\, dx = \int_{x}^{\infty} f(x)\, dx = .5$$

probability distribution, probability function or **probability mass function,** *n.* the distribution of the probabilities of the different values of a discrete random variable X.

Example: the probability distribution of the number of girls in families of 3 children, is as follows:

X (no. of girls)	0	1	2	3
Probability p	$\frac{1}{8}$	$\frac{3}{8}$	$\frac{3}{8}$	$\frac{1}{8}$

If the variable X takes values $x_1, x_2, \ldots x_n$, with probabilities $p(x_1), p(x_2) \ldots p(x_n)$, its probability distribution must satisfy two conditions:

(a) The sum of all the probabilities is 1;

$$\sum_i p(x_i) = 1$$

(b) No probability can be negative;

$$p(x_i) \geqslant 0 \quad \text{all i}$$

The mean (or expectation) and variance of X are given by:

$$Mean = \sum_i x_i p(x_i)$$

$$Variance = \sum_i x_i^2 p(x_i) - \text{mean}^2$$

The probability distribution for the BINOMIAL DISTRIBUTION is given by:

$$p(X=r) = {}^nC_r p^r (1-p)^{n-r}$$

where $0 \leqslant r \leqslant n$.

The probability distribution for the POISSON DISTRIBUTION is:

$$p(X=r) = e^{-\mu}\frac{\mu^r}{r!} \qquad \text{where } r \geqslant 0.$$

For continuous distributions, the equivalent of probability distribution is PROBABILITY DENSITY FUNCTION.

probability function, see PROBABILITY DISTRIBUTION.

probability-generating function, *n.* an expression which, when expanded, has terms whose coefficients are the probabilities of the different numbers of occurrences of an event.

Example:

$$\left(\tfrac{1}{2}+\tfrac{1}{2}t\right)^3$$

can be expanded to give

$$\tfrac{1}{8}+\tfrac{3}{8}t+\tfrac{3}{8}t^2+\tfrac{1}{8}t^3$$

representing the probabilities of $\frac{1}{8}, \frac{3}{8}, \frac{3}{8}$ and $\frac{1}{8}$ of 0, 1, 2 and 3 tails respectively when a coin is tossed three times.

The probability-generating function for the BINOMIAL DISTRIBUTION with probability p of success, $q=1-p$ of failure, is given for n trials by:

$$(q+pt)^n$$

where t denotes a success.

The probability-generating function for the POISSON DISTRIBUTION is given by:

$$\sum_{r=0}^{\infty} e^{-\mu}\frac{(\mu t)^r}{r!}$$

where $\mu=$ mean.

When a single die is thrown, the probability-generating function for the score shown is

$$\frac{t}{6}+\frac{t^2}{6}+\frac{t^3}{6}+\frac{t^4}{6}+\frac{t^5}{6}+\frac{t^6}{6}$$

That is, all the scores 1 to 6 have probabilities 1/6. When two dice are thrown, the probability-generating function for the total scores, from 2 to 12, is:

$$\left(\frac{t}{6}+\frac{t^2}{6}+\frac{t^3}{6}+\frac{t^4}{6}+\frac{t^5}{6}+\frac{t^6}{6}\right)^2$$

If the expression were raised to the power 3, it would give the probabilities of the outcomes of the total of three dice. And so on.

If G(t) is the probability-generating function of a probability distribution *X*, then

$$G(1) = 1$$
$$G'(1) = E(X)$$

($G'(1)$ means the value of $\frac{dG}{dt}$ when $t = 1$, etc.).

$$\text{Var}(X) = G''(1) + G'(1) - [G'(1)]^2$$

probability mass function, see PROBABILITY DISTRIBUTION.

process average, *n.* in quality control, the proportion of defective items produced by a machine or manufacturing process when it is running normally.

producer's risk, *n.* in quality control, the probability that a batch with proportion of defectives equal to the PROCESS AVERAGE (i.e., an average batch) will be rejected by the inspection scheme.

product-moment correlation, see CORRELATION COEFFICIENT.

proportions distribution, *n.* the distribution of the proportion of successes in a set of Bernoulli trials. If each trial has probability p of success, $q = (1 - p)$ of failure, this distribution has

$$\textit{Mean} = p$$

$$\textit{Standard deviation} = \sqrt{\frac{pq}{n}}$$

Q

quadrat, *n.* an area, usually of vegetation, randomly selected for study. It is normally square in shape (hence the name). The ideal size of a quadrat is the smallest size that contains the same number of species as would be contained in a larger one. In a *point quadrat,* sampling is carried out at the points of a square grid covering the quadrat.

When several quadrats are placed in a row, the area formed is called a *transect* (Fig. 91).

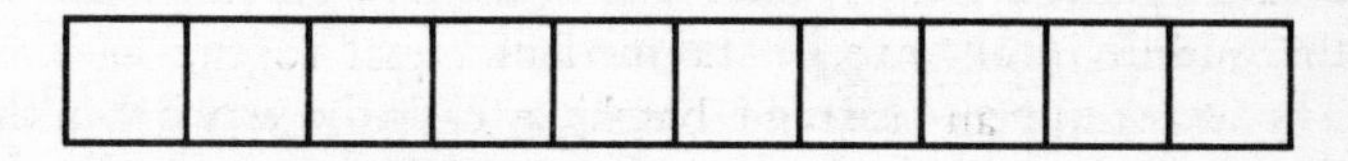

Fig. 91. **Quadrat.** Transect comprising 10 quadrats.

qualitative data, *n.* data which is classified by type. Compare QUANTITATIVE DATA. Thus data on how hospital patients feel after a new course of treatment can be grouped qualitatively as:

Much better Better The same Worse Much worse.

If however, their number of hours sleep is recorded, this data is quantative, being expressed numerically.

quality control, *n.* control of the relative quality of a manufactured product, usually by statistical sampling techniques.

quantile, *n.* general name for the values of a variable which divide its distribution into equal groups. A set of values may be ranked according to size. Thus 6, 8, 4, 3, 9, would be ranked as:

Number	*Rank*
9	1
8	2
6	3
4	4
3	5

(The ranking used to find quantiles is slightly different from that used at other times; when a tie occurs, the values are ranked one above the other, rather than equal.)

When a set of n values is ranked in this way, it may be divided into four equal groups by the quartiles, into six equal groups, when the divisions occur at the sextiles, into eight, with divisions at the octiles, ten at the deciles, and a hundred the percentiles.

The MEDIAN or second QUARTILE has the middle position. In the set of five numbers given above, the median is the third number, namely 6. For n numbers, the median is that with position $\frac{1}{2}(n+1)$ in the ranked order. If n is even there is no actual middle member; the mean of the two on either side of the middle is taken to be the median.

The 1st, or upper, quartile is one-quarter of the way down the distribution at rank number $\frac{1}{4}(n+1)$. The 3rd, or lower, quartile is three-quarters of the way down, rank number $\frac{3}{4}(n+1)$. Thus the three quartiles divide the set into four equal groups.

quantitative data, *n.* data which is classified by some numerical value. Compare QUALITATIVE DATA. Thus, data on the number of hours of sleep recorded for certain hospital patients is expressed numerically, and is quantitative data.

quartile, *n.* the value of a variable below which three quarters (1st or upper quartile) or one quarter (3rd or lower quartile) of a distribution lie. See QUANTILES. The median is the second quartile. See also INTERQUARTILE RANGE, DISPERSION, EXPLORATORY DATA ANALYSIS.

quartile coefficient of dispersion, see ABSOLUTE MEASURE OF DISPERSION.

quartile coefficient of skewness, see SKEW.

quartile deviation, see INTERQUARTILE RANGE.

questionnaire, *n.* a set of questions on a form submitted to a number of people in order to collect information.

quota sampling, *n.* a sampling scheme often used in market and social surveys in which one or more interviewers are used, each being given instructions about the section of the population which is his or her responsibility. The actual choice of interviewees is left to the interviewer. Each of them might, for example, be told to select 20 adult women, 20 adult men, 10 teenage girls, 10 teenage boys and 10 younger children from their home town.

R

random error, *n.* the deviation of an observed value from its expected value due to the inherent randomness of the situation under study. See ERROR.

randomized blocks design, see EXPERIMENTAL DESIGN.

random numbers, *n.* numbers selected at random; that is, each number of a set has an equal chance of being selected, and each selection is independent of all previous selections.

Random numbers are used in the design and implementation of random sampling schemes. They are also used in SIMULATION; for example, a flight simulator for training pilots uses random numbers to generate gusts and changes of direction to be imposed on a steady basic wind.

Random numbers are given in tables, such as those on page 267. Each of these numbers is selected at random from the digits 0, 1, 2, 3, 4, 5, 6, 7, 8 and 9. Almost all computers have a random number generating facility.

random sampling, *n.* sampling in which every individual has a known probability of being selected. Examples of random sampling are SIMPLE RANDOM SAMPLING, CLUSTER SAMPLING, QUOTA SAMPLING, STRATIFIED SAMPLING and SYSTEMATIC SAMPLING.

random selection, *n.* selection in which all items have a known probability of being selected. See, for example, SIMPLE RANDOM SAMPLING.

random variable, *n.* a variable which takes values in a certain range with probabilities specified by a PROBABILITY DISTRIBUTION or PROBABILITY DENSITY FUNCTION.

When a coin is tossed, the outcome is a head or a tail. This is not a random variable because it is not a number. If, however, the result is expressed as the number of tails, it is a random variable which can take the values 0 or 1, each with probability $\frac{1}{2}$.

range, *n.* the difference between the largest and the smallest values of a sample or set of variables (see DISPERSION). The range of 65,63,58,52,47 is $65-47=18$.

The STANDARD DEVIATION of a normally-distributed population may be estimated from the range of a sample, using the table on page 257.

rank, *n.* the position, when in order, of a member within a set. When two or more members are equal, they are often given rank equal to the mean of the ranks they would have had if they had been slightly different. In that case they are said to be tied. Example: rank the numbers 47, 50, 49, 65, 68, 49, 49, 65.

Number	*Rank*	
68	1	
65	$2\frac{1}{2}$	Mean of 2 and 3 is $2\frac{1}{2}$
65	$2\frac{1}{2}$	
50	4	
49	6	
49	6	Mean of 5, 6 and 7 is 6
49	6	
47	8	

For most tests involving ranks it does not matter if the rank starts with the highest or the lowest number as rank 1, so long as the test is consistent throughout. This is not true, however, of the WILCOXON MATCHED-PAIRS SIGNED-RANK TEST, where the smallest difference must count as rank 1. See also QUANTILES.

rank correlation, *n.* correlation of data by rank, rather than by numerical value. This can be measured using Kendall's or Spearman's rank correlation coefficient, see CORRELATION COEFFICIENT.

ratio scale, *n.* a scale of measurement in which equal differences between points correspond to equal distances on the scale, and there is a true zero.

An example of a ratio scale is length. The difference between 2m and 2.50m is the same as that between 10.25m and 10.75m.

A length of zero means no length. Consequently it is possible to speak of ratios of lengths. A length of 6m is twice as long as one of 3m.

rectangular distribution or **uniform distribution,** *n.* a continuous distribution which is constant over a certain range of values, and zero elsewhere. Plotting the probability density function f(x) produces a rectangular-shaped graph (Fig. 92).

$$\textit{Mean} = \frac{a+b}{2}$$

$$\textit{Variance} = \tfrac{1}{12}(b-a)^2$$

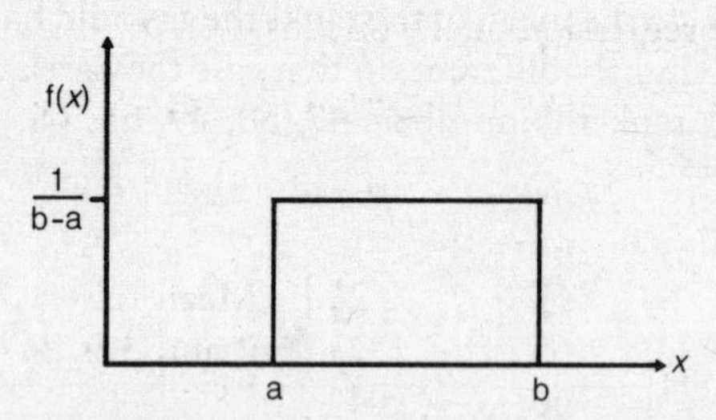

Fig. 92. **Rectangular distribution** or **uniform distribution.** In this example, f(x) = 1/(b − a), where a < x < b f(x) = 0 elsewhere. Thus the area under the graph is 1, as required.

regression line, *n.* a LINE OF BEST FIT drawn through a set of points on a graph. When y is the DEPENDENT VARIABLE and x the independent, the regression line (y on x) is given by:

$$y - \bar{y} = \frac{s_{xy}}{s_x^2}(x - \bar{x})$$

Example: find the regression lines through the points (2, 3), (4, 5), (6, 6), (5, 8) and (8, 13).

x	y	$x-\bar{x}$	$y-\bar{y}$	$(x-\bar{x})(y-\bar{y})$	$(x-\bar{x})^2$	$(y-\bar{y})^2$
2	3	−3	−4	12	9	16
4	5	−1	−2	2	1	4
6	6	1	−1	−1	1	1
5	8	0	1	0	0	1
8	13	3	6	18	9	36
25	35			35	20	58

$$\bar{x} = \frac{25}{5} = 5 \qquad \bar{y} = \frac{35}{5} = 7$$

$$s_{xy} = \frac{\sum (x-\bar{x})(y-\bar{y})}{n} = \frac{35}{5} = 7$$

$$s_x^2 = \frac{\sum (x-\bar{x})^2}{n} = \frac{20}{5} = 4$$

$$s_y^2 = \frac{\sum (y-\bar{y})^2}{n} = \frac{58}{5} = 11.6$$

Thus the regression line, y on x, is

$$y - 7 = \tfrac{7}{4}(x-5)$$
$$y = 1.75x - 1.75$$

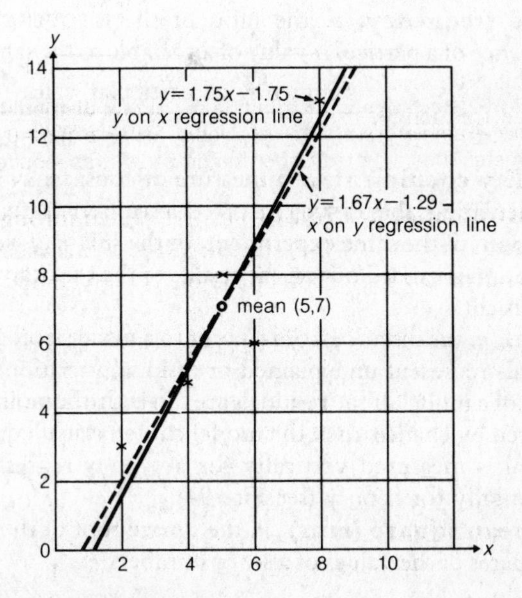

Fig. 93. **Regression line.** The x on y, and the y on x regression lines, and the mean for the points (2,3), (4,5), (6,6), (5,8) and (8,13).

and the other regression line, x on y, is

$$x - 5 = \frac{7}{11.6}(x - 7)$$

$$x = .6y + .775$$

$$(\text{or } y = 1.67x - 1.29)$$

The two regression lines (see Fig. 93) always meet at the mean of the points $(\bar{x}, \bar{y})$. This will not usually be one of the original points.

The regression line formula is proved by the method of least squares (see LEAST SQUARES, METHOD OF) which minimizes the sum of the squares of the RESIDUALS, the distances from the points to the regression line measured vertically (y on x) or horizontally (x on y).

relative frequency, *n.* the ratio of the FREQUENCY of the occurrence of a particular value of a variable to the sample size.

$$\text{Relative frequency} = \frac{\text{frequency of particular value (or class)}}{\text{total sample size}}$$

reliability coefficient, *n.* a measure of consistency obtained by calculating the CORRELATION COEFFICIENT between two repetitions of the same experiment. In the *split-half method*, the correlation is carried out on the results of the two halves of one experiment.

residual, *n.* the distance from a point to a REGRESSION LINE. The residuals represent unexplained or residual variation after the fitting of a model. If the residuals are too big to have reasonably occurred by chance, then the model chosen is inadequate. The residual is measured vertically for a y on x regression line, horizontally for x on y (see Fig. 94).

root mean square (rms), *n.* the square root of the mean of the squares of the values of a set of numbers.

$$\text{rms} = \sqrt{\left(\frac{\sum_i x_i^2}{n}\right)}$$

If the mean of the numbers is zero (as is the case when they represent deviations from their mean) the root mean square is the same as the STANDARD DEVIATION.

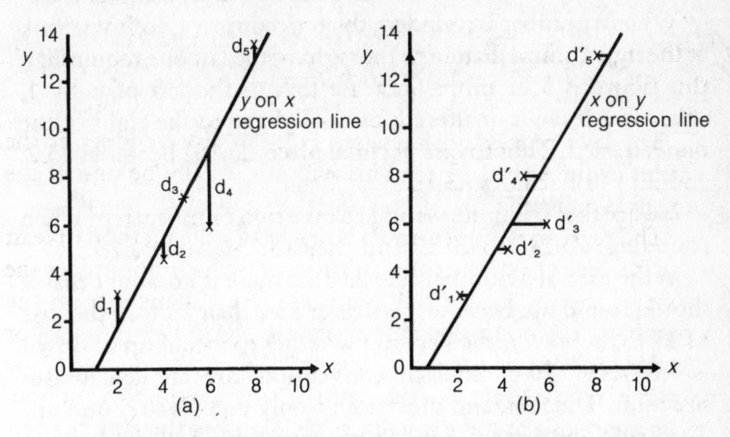

Fig. 94. **Residual.** (a) Residuals $d_1, d_2, \ldots d_5$, for the *y* on *x* regression line. (b). Residuals $d_1', d_2', \ldots d_5'$, for the *x* on *y* regression line.

root mean square variation, *n.* the square root of the mean of the squares of the deviations of a set of numbers from an assumed mean. For an assumed mean *a*, root mean squared deviation is given by:

$$\sqrt{\frac{\sum (x-a)^2}{\text{n}}}$$

Example: the root mean square variation of 5, 4, 4, 3, 2, and 0 about 4 is:

$$\sqrt{\frac{[1^2+0^2+0^2+(-1)^2+(-2)^2+(-4)^2]}{6}}=1.9$$

When the assumed mean a is the same as the true mean $\bar{x}$, the root mean square variation is the same as the STANDARD DEVIATION.

rounding, *vb.* adjusting a number, up or down, to a required level of accuracy. There are several ways in which the level of accuracy may be presented, such as to a given number of decimal places, to a given number of significant figures, to the nearest whole number, to the nearest hundred, thousand, etc.

When a number is rounded, the procedure is as follows: look at the figure immediately to the right of the last one required; if this figure is 5 or more, increase that to the left of it by 1, otherwise leave it unaltered; omit all digits to the right of the one required. Thus to one decimal place, 23.152 becomes 23.2, and 23.148 becomes 23.1.

Notice that it is quite wrong to take two or more steps when rounding; 23.148 to 23.15 and then to 23.2 is incorrect.

In the case of a number like 23.152, there is no doubt that it should round up, because it is nearer 23.2 than 23.1. In the case of 23.15, however, the decision whether to round up or down is arbitrary. Two different conventions are applied in this situation. The first, and most commonly used, is to round up. This rule, however, applied to statistical data leads to a systematic error. For example, adding 23.15, 23.25, 23.35 and 23.45 gives a total of 93.20. If these figures are each rounded to one decimal place, and added, an error is produced:

$$23.2 + 23.3 + 23.4 + 23.5 = 93.4$$

To overcome this, a rule is sometimes used that in this type of situation, one rounds up if the preceding number is odd, down if it is even. Thus,

23.15 is rounded up, 23.2
23.25 is rounded down, 23.2
23.35 is rounded up, 23.4
23.45 is rounded down, 23.4

When these new figures are added, the total is $93.\dot{2}$.

S

sample, *n.* a set of individuals or items selected from a PARENT POPULATION so that properties or parameters of the population may be estimated, or so that hypotheses about the population may be tested. The distribution of the means of samples of a particular size is called the sampling distribution of the means; similarly the distribution of the variances of the samples is the sampling distribution of the variances; and so on.

If sampling without replacement, with sample size n from a parent population size N, mean μ and standard deviation σ, then the sampling distribution of the means has

$$\textit{Mean} = \mu$$

$$\textit{Standard deviation} = \frac{\sigma}{\sqrt{n}}\sqrt{\frac{N-n}{N-1}}$$

and is normally distributed (or approximately so) if the sample size is large.

If the size of the parent population is large, or sampling with replacement, then the size of the parent population is effectively infinite; the standard deviation of this sampling distribution can be simplified to:

$$\frac{\sigma}{\sqrt{n}}$$

which is the STANDARD ERROR, and is the standard deviation of the means of samples of size n.

If a sample of size n has mean m and standard deviation s, then unbiased estimators for the parent population are:

$$\textit{Parent mean} = \mathrm{m}$$

$$\textit{Parent variance} = \mathrm{s}^2\left(\frac{\mathrm{n}}{\mathrm{n}-1}\right)$$

Example: the yields of 10 potato plants (in kg) are 1.4, 1.2, 1.3, 1.0, 1.8, 2.4, 1.6, 0.8, 1.1 and 1.4. For these figures,

Sample mean = 1.4
Sample standard deviation = .431
Estimated parent mean = 1.4

$$\textit{Estimated parent standard deviation} = .431\sqrt{\frac{10}{9}} = .454$$

If two samples (from the same parent population) have sizes n_1 and n_2, means m_1 and m_2, and standard deviations s_1 and s_2, then,

$$\textit{Estimated parent mean} = \frac{m_1 n_1 + m_2 n_2}{n_1 + n_2}$$

$$\textit{Estimated parent standard deviation} = \sqrt{\frac{(n_1 s_1^2 + n_2 s_2^2}{(n_1 + n_2 - 2)}}$$

For k such samples,

$$\textit{Estimated parent mean} = \frac{m_1 n_1 + m_2 n_2 + \ldots + m_k n_k}{n_1 + n_2 + \ldots + n_k}$$

Estimated parent standard deviation

$$= \sqrt{\frac{(n_1 s_1^2 + n_2 s_2^2 + \ldots + n_k s_k^2)}{(n_1 + n_2 + \ldots + n_k - k)}}$$

sample point, *n.* an element in a SAMPLE SPACE.

sample size, *n.* the number of individuals or items in a sample. Samples are often called small if their size is under about 30.

sample space, *n.* the set of individuals or items available to be sampled; the PARENT POPULATION from which a sample is drawn. See also OUTCOME.

sampling distribution, *n.* the distribution of a statistic of a SAMPLE of a particular size from a given PARENT POPULATION; frequently the statistic in question is the mean. The sampling distribution of the means of samples of size n is normal, or approximately so (if the sample size is large enough), with mean equal to the parent mean μ, and the standard deviation given by:

$$\frac{\sigma}{\sqrt{n}}\sqrt{\frac{N-n}{N-1}}$$

where N is the size of the parent population. If N is large, or the sampling is done with replacement, this can be simplified to:

$$\frac{\sigma}{\sqrt{n}}$$

If two samples are taken from different parent populations, with means μ_1 and μ_2, and standard deviations σ_1 and σ_2, the sampling distribution of the difference in their means has

$$\textit{Mean} = \mu_1 - \mu_2$$

$$\textit{Standard deviation} = \sqrt{\frac{\sigma_1^2}{n_1} + \frac{\sigma_2^2}{n_2}}$$

If two samples are taken from the same parent population, the sampling distribution of the difference in their means has

$$\textit{Mean} = 0$$

$$\textit{Standard deviation} = \sqrt{\frac{\sigma^2}{n_1} + \frac{\sigma^2}{n_2}}$$

sampling error, *n.* the difference between the true value of a parameter of a parent population and that estimated from a sample. This error is due to the fact that the value has been calculated from a sample rather than from the whole parent population.

Example: a large shoal of fish has mean mass 0.92kg, with standard deviation .25kg. A boy catches five fish from the shoal with masses .81, .63, 1.04, 1.11 and .91kg. The mean mass of his fish is .90kg and he concludes this is the mean mass of the shoal. The sampling error for the mean is:

$$.92 - .90 = .02\text{kg}$$

He estimates the standard deviation of the shoal to be .19kg using

$$\text{Parent SD} = \text{Sample SD}\sqrt{\frac{n}{n-1}}$$

as an ESTIMATOR. The sampling error for the standard deviation is thus

$$.25 - .19 = .06\text{kg}$$

sampling fraction, *n.* at its simplest, sample size divided by the size of parent population. This definition cannot, however, be applied to all sampling schemes. In STRATIFIED SAMPLING, for example, each stratum has its own sampling fraction.

sampling line, *n.* the line on a LATTICE DIAGRAM illustrating good and defective samples.

scatter diagram, *n.* a Cartesian GRAPH illustrating a sample from a BIVARIATE DISTRIBUTION. The points are plotted but not joined. See also FREQUENCY SURFACE.

A scatter diagram is usually drawn before working out a CORRELATION COEFFICIENT; it gives a good initial indication of whether there is any point in doing this calculation. If the points lie roughly on a straight line, it is reasonable to go ahead, but if they are scattered all over the graph paper, the calculation is likely to be a waste of time.

A scatter diagram will also show up a non-linear relationship (which may give a low value of product-moment correlation coefficient, Fig. 95), OUTLIERS, and a BIMODAL distribution (which can give a deceptively large product-moment correlation coefficient).

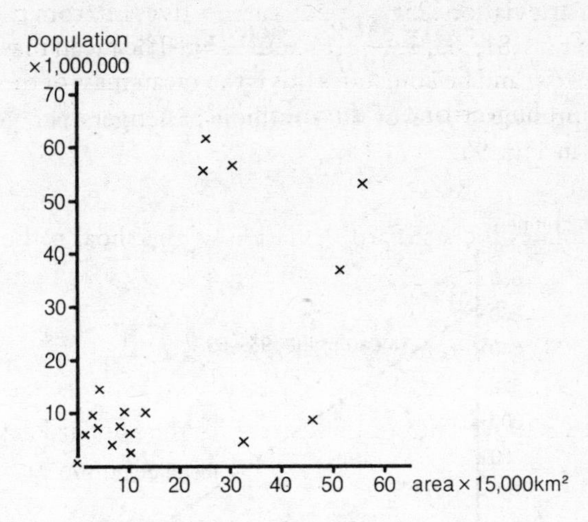

Fig. 95. **Scatter diagram.** Population against land area for western European countries (1976). Source: *Philip's Certificate Atlas*, 1978.

seasonal component, *n.* the component of variation in a TIME SERIES which is dependent on the time of year. Symbol: S. The costs of various types of fruit and vegetable, unemployment figures, and mean daily rainfall, all show marked seasonal variation. Compare IRREGULAR VARIATION.

semi-averages, method of, *n.* a method for estimating TREND. The data are in the form of a TIME SERIES; the figures are divided chronologically into two groups, and the means of each taken. These means are then used to estimate the trend. Example: the passenger journeys on British Rail (millions) between 1965 and 1974 were as follows (Sources: British Rail, H.M.S.O. *Annual Abstract of Statistics, 1975*):

Year	1965	1966	1967	1968	1969	1970	1971	1972	1973	1974
Journeys	865.1	835.0	837.4	831.1	805.2	823.9	815.5	753.6	728.0	719.0

The mean for the first 5-year period (1965–1969) = 834.8
The mean for the second 5-year period (1970–1974) = 768.0
The means of the two sets of years are 1967 and 1972.

So the trend is estimated as:

$$\frac{768.0-834.8}{1972-1967}=\frac{-66.8}{5}=-13.4$$

This represents a loss of 13.4 million passengers per year, as shown in Fig. 96.

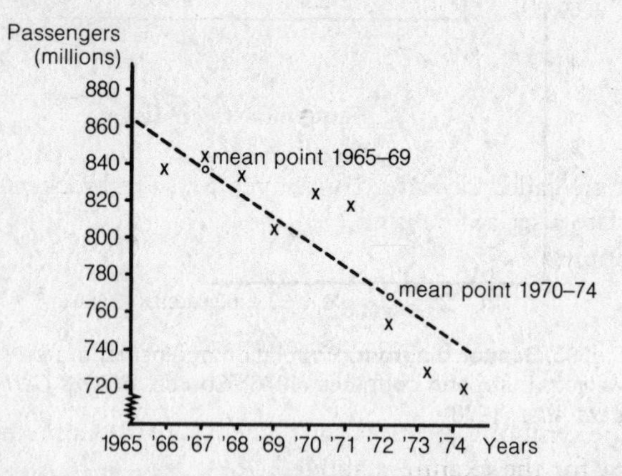

Fig. 96. **Semi-averages, method of.** Graph illustrating the trend of passenger journeys on British Rail between 1965 and 1974.

semi-interquartile range, *n.* a measure of DISPERSION given by:

$$\frac{\text{upper quartile} - \text{lower quartile}}{2}$$

See also INTERQUARTILE RANGE.

semi-logarithmic graph, see GRAPH.

sequential sample, *n.* in quality control, a sampling scheme in which the lower line of a LATTICE DIAGRAM goes upwards in a series of steps, and the sample is continued until the sampling line enters either the acceptance or rejection region (Fig. 97, opposite).

set, *n.* a collection of numbers, objects, etc., that usually have at least one common property or characteristic. The members of

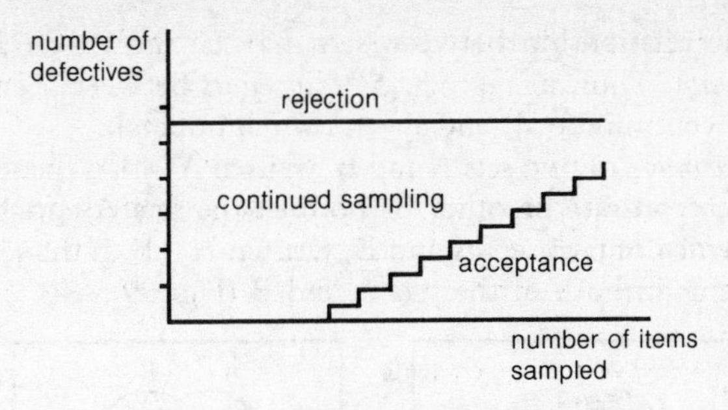

Fig. 97. **Sequential sample.**

a set are called *elements*. By convention, the brackets used to indicate a set are written { }.
Examples:

$$V = \{a, e, i, o, u, y\}$$
$$P = \{\text{prime numbers} < 30\}$$
$$M = \{\text{Malaysian states}\}$$

The symbol ∈ means is 'an element of' or 'is a member of'. Thus, for the examples above,

u ∈ V   17 ∈ P   Trengganu ∈ M

The set from which members of a particular set may be drawn is called the *universal set*, denoted by ℰ. Thus, in the case of V = {a,e,i,o,u,y} ℰ could be {all letters of the alphabet}; in the case of P = {prime numbers < 30}, ℰ could be {integers between 1 and 30}.

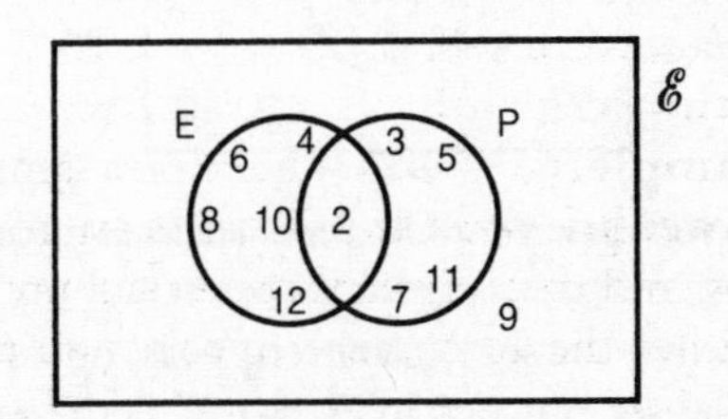

Fig. 98. **Set.** A Venn diagram where ℰ is {integers between 2 and 12}, E is {even numbers} and P is {prime numbers}.

The relationship between sets may be shown on a VENN DIAGRAM. Thus, in Fig. 98, $\mathscr{E}$ = {integers between 2 and 12}, E = {even numbers}, and P = {prime numbers}.

The *union* of two sets A and B, written $A \cup B$, is the set of all elements in one or other or both of the sets A and B. The *intersection* of two sets A and B, written $A \cap B$, is the set of all elements in both of the sets A and B (Fig. 99).

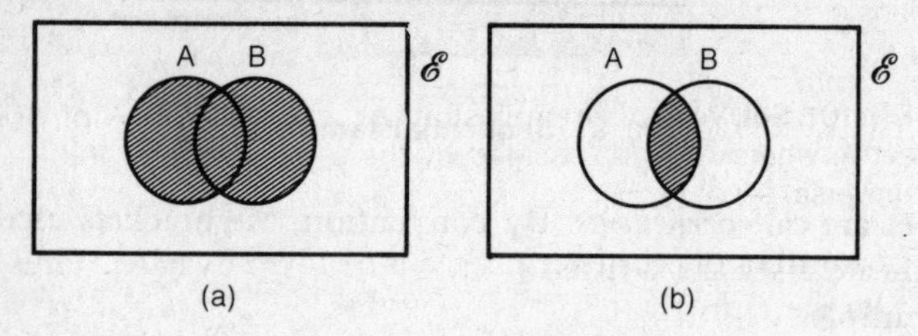

Fig. 99. **Set.** (a) $A \cup B$ (b) $A \cap B$.

Example:

$$A = \{l,a,m,p\}$$
$$B = \{s,h,a,d,e\}$$
$$A \cup B = \{l,a,m,p.s,h,d,e\}$$
$$A \cap B = \{a\}$$

If every element of a set A also belongs to set B, then A is said to be a *subset* of B; this is written $A \subset B$ (see Fig. 100).

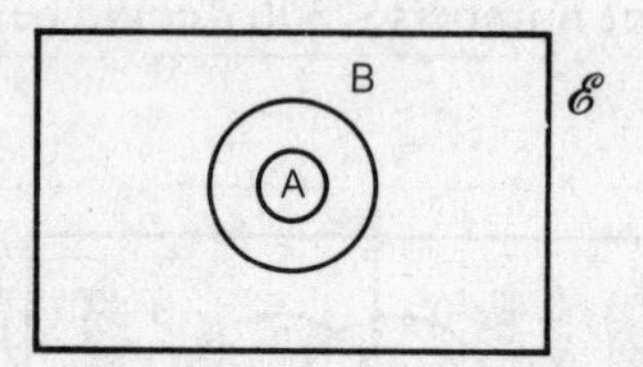

Fig. 100. **Set.** Venn diagram illustrating $A \subset B$.

The *complement* of a set A, written A′, is the set of all elements in the universal set but not in A. So A′ is the set {not A}. For example, if $\mathscr{E}$ = {1,2,3,4,5,6} and A is {1,2,3,4} then A′ = {5,6} as in Fig. 101.

The set with no members is called the *empty set* and written $\emptyset$ or $\{\}$.

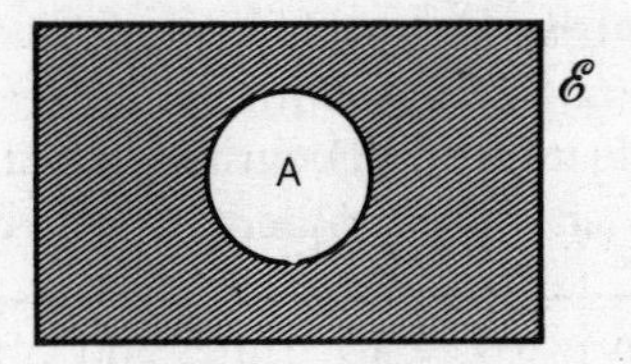

Fig. 101. **Set.** Venn diagram illustrating the complement A′ of set A, where A′ is represented by the shaded area of the universal set ℰ.

The number of elements in set A is denoted by n(A). Thus, if $A=\{a,n,g,e,r\}$, $n(A)=5$.
For two sets A and B,

$$n(A)+n(B)=n(A\cap B)+n(A\cup B)$$

The translation of set language into English is illustrated in the example, (Fig. 102) where $\mathscr{E}=\{\text{Animals}\}$, $C=\{\text{Cats}\}$, $D=\{\text{Dogs}\}$, $S=\{\text{Siamese cats}\}$, and $T=\{\text{Tame animals}\}$.

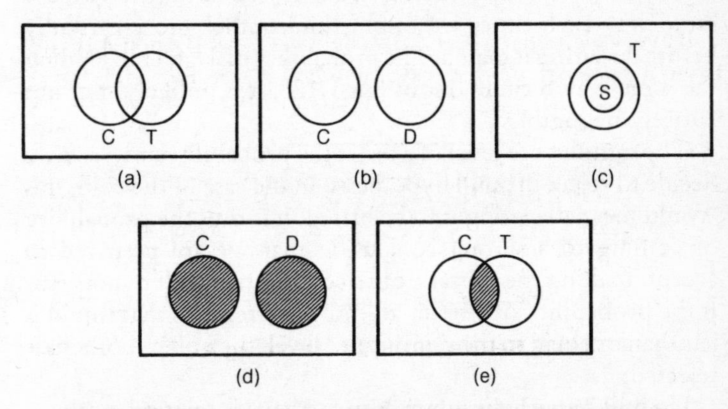

Fig. 102. **Set.** (a) $C\cap T\neq\emptyset$; some cats are tame. (b) $C\cap D=\emptyset$; no cats are dogs. (c) $S\subset T$; all Siamese cats are tame. (d) $C\cup D$; animals which are either cats or dogs. (e) $C\cap T$; animals which are both cats and tame, i.e. tame cats.

Sheppard's correction, see GROUPED DATA.

sigma (Σ) notation, see SUMMATION NOTATION.

significance level, *n.* the probability of rejecting a true NULL HYPOTHESIS in a statistical test.

It is not possible to draw conclusions from statistical data in the way that one can, for example, prove that the angle sum of a triangle is 180°, i.e. with absolute certainty.

Suppose a coin is tossed 100 times and comes down heads every time. There is clearly strong reason to suspect the coin of being biased, but there is still the remote possibility that it is unbiased and just happened to fall that way. What can be said is that the probability of a fair coin landing the same way 100 times out of 100 is $2 \times (\frac{1}{2})^{100}$, which is 1.6×10^{-30}. Thus if, on this evidence, the null hypothesis, 'The coin is unbiased' is rejected in favour of the alternative hypothesis, 'The coin is biased', there is a very small probability that the conclusion is incorrect.

In that case, the probability of error is clearly so small that we would have no hesitation in declaring the coin biased, even though that conclusion could be wrong, theoretically. If, however, the coin had been tossed 8 times and came down the same way each time, we might think rather more carefully before declaring it biased. The probability of a fair coin landing the same way 8 times out of 8 is 1/128; it is unlikely, but not entirely negligible.

The significance level of a test is the probability at which we decide to reject the null hypothesis. In the case of the coin, this would mean declaring it to be biased. It is thus the probability of coming to a wrong conclusion that we are prepared to accept, making the mistake of rejecting a true null hypothesis. If the probability of a result at least as extreme as that found is less than or equal to the significance level, the null hypothesis is rejected.

If a null hypothesis which is in fact true is rejected, a type 1 error occurs (see TYPE 1 AND TYPE 2 ERRORS). If a false null hypothesis is accepted, a type 2 error occurs. The probabilities of each of these types of error are related. The lower the significance level of the test, the lower is the probability of a type 1 error, but the greater that of a type 2 error. The significance level chosen for any particular test depends on the consequences of each type of error.

Example: a sample of the rivets on the wing of an aircraft is tested for possible weakening.

Null hypothesis: the rivets are of the correct strength.

Alternative hypothesis: the rivets have become weaker.

Type 1 error: the rivets are all right, but declared to be weak.

Type 2 error: the rivets are weak, but declared to be all right.

The consequence of a type 1 error would be that expensive maintenance work is carried out unnecessarily.

The consequence of a type 2 error would be that the aircraft's safety is in danger.

In this case a type 1 error is clearly more desirable (or less undesirable) than a type 2 error, and so the significance level would be set at quite a high figure.

Statistical tests are usually carried out at predetermined significance levels. Typical statements of results might be: 'There is no reason, at the 10% significance level, to reject the null hypothesis that there is no difference in height between the adult men in Accra and Lagos'. 'The claim that the machine is producing nails shorter than specified is accepted at the 5% significance level: the null hypothesis that their length is correct is rejected'.

In many tests the probability need not actually be worked out. The test data are used to work out a statistic, like χ^2 or t, which is compared to a critical value found in tables (see p. 249, p. 248), for the required significance level and number of degrees of freedom.

significant figures, *n.* **1.** the figures of a number that express its magnitude to a specified degree of accuracy. **2.** the number of such figures.

Example: 3.141 592 7 written to 4 significant figures is 3.142. The digits, 3, 1, 4 and 2, are the four significant figures.

sign test, *n.* a non-parametric test of the NULL HYPOTHESIS, that two matched samples are drawn from the same population. It actually tests for a MEDIAN difference of zero.

Example: ten hospital patients were given a new sleeping pill. The number of hours they slept that night were recorded and compared to the figures for the previous night, when they had been on the old pill. Is there reason, at the 5% SIGNIFICANCE LEVEL, to think the new pill different from the old one in its effect?

Null hypothesis: There is no difference in the effects of the two pills.

Alternative hypothesis: There is a difference.

	Old pill	*New pill*	*Sign of difference*
Miss Pink	8	$8\frac{1}{4}$	+
Mr Black	6	$4\frac{1}{2}$	−
Mrs Green	7	7	0
Mrs Brown	5	9	+
Mr Purple	7	$7\frac{1}{2}$	+
Miss White	$6\frac{1}{2}$	$6\frac{1}{4}$	−
Ms Blue	8	8	0
Mr Carmine	7	$7\frac{1}{2}$	+
Mrs Red	4	5	+
Mr Orange	6	$8\frac{1}{4}$	+

In the right hand column, sign of difference, those who slept longer have been given +, shorter −, and those who slept for the same hours, 0.

The two who slept the same length of time, Mrs Green and Ms Blue, are discounted from the sample. Thus there remain 8 members, 6 + and 2 −.

The null hypothesis can now be restated as the probability of a positive result + equals the probability of a negative result −, which equals $\frac{1}{2}$.

It is now possible to apply the BINOMIAL DISTRIBUTION to this situation (see also PASCAL'S TRIANGLE). If p means a + result, m a − result, the probabilities are generated by $(\frac{1}{2}p + \frac{1}{2}m)^8$. When this is expanded, it gives the probabilities of the possible outcomes as follows:

$$(\tfrac{1}{2}p)^8 + 8(\tfrac{1}{2}p)^7(\tfrac{1}{2}m) + 28(\tfrac{1}{2}p)^6(\tfrac{1}{2}m)^2 + 56(\tfrac{1}{2}p)^5(\tfrac{1}{2}m)^3$$
$$+ 70(\tfrac{1}{2}p)^4(\tfrac{1}{2}m)^4 + 56(\tfrac{1}{2}p)^3(\tfrac{1}{2}m)^5 + 28(\tfrac{1}{2}p)^2(\tfrac{1}{2}m)^6$$
$$+ 8(\tfrac{1}{2}p)(\tfrac{1}{2}m)^7 + (\tfrac{1}{2}m)^8$$

which can be summarized as:

Outcome	8+, 0−	7+, 1−	6+, 2−	5+, 3−	4+, 4−
Probability	$\frac{1}{2^8}$	$\frac{8}{2^8}$	$\frac{28}{2^8}$	$\frac{56}{2^8}$	$\frac{70}{2^8}$

Outcome	0+, 8−	3+, 5−	2+, 6−	1+, 7−
Probability	$\frac{56}{2^8}$	$\frac{28}{2^8}$	$\frac{8}{2^8}$	$\frac{1}{2^8}$

The actual result obtained was 6+, 2−. The probability of a result as far or further from the central value of 4+, 4− as this, is that of 8+, 0−, or 7+, 1−, or 6+, 2−, or 2+, 6−, or 1+, 7−, or 0+, 8−, namely:

$$\frac{1}{2^8}+\frac{8}{2^8}+\frac{28}{2^8}+\frac{28}{2^8}+\frac{8}{2^8}+\frac{1}{2^8}=\frac{74}{256}=.29$$

This is not a low probability, and so there is no reason to reject the null hypothesis. To do so, at the 5% level, the probability would have to be less than 5%, or .05. The new pill, therefore, has not proved itself.

If cumulative probability tables for the binomial distribution (or $p = 1/2$) are available, the calculation can be largely avoided.

The sign test is very crude; in the example used, it can be seen that no account was taken of how much longer, or shorter, the various patients slept. Miss Pink slept only $\frac{1}{4}$ hour more, Mrs Brown 4 hours, but both have the same rating as a single +. It is thus not particularly well-suited for this sort of situation, involving quantitative data. A WILCOXON MATCHED-PAIRS SIGNED-RANK TEST or a paired sample t-TEST would have been more appropriate. Had the patients, on the other hand, been asked a qualitative question like, 'Was your sleep more relaxed?', the sign test would have been quite suitable, counting + for 'Yes', − for 'No', and 0 for 'No change'.

simple random sampling, *n.* sampling in which each member of the parent population has equal probability of being selected, and each selection is independent of all others. Equal probability can be true of other methods of sampling, for example, SYSTEMATIC SAMPLING with a random start, and STRATIFIED SAMPLING with subsample sizes proportional to the size of the strata. In both of these cases there are many samples of the parent population which could not be obtained because the items are not selected independently of one another. In simple random sampling every possible sample of a given size has the same probability of selection. Compare CLUSTER SAMPLING.

simulation, *n.* the use of a mathematical model to reproduce the conditions of a situation or process, and to carry out experiments on it. Simulation may be carried out for reasons of economy, or because it is difficult (or impossible) to conduct an EXPERIMENT.

Simulation of statistical processes or experiments is often done by computer using RANDOM NUMBERS.

single sampling scheme, *n.* in quality control, a sampling scheme where only one sample per batch is inspected. If the number of defectives found is not more than the ACCEPTANCE NUMBER, the batch is accepted. Otherwise it is rejected, or subjected to 100% inspection.

skew, *adj.* (of a distribution) not having equal probabilities above and below the MEAN. In a skew distribution, the median and the mean are not coincident. If the median is less than the mean, the distribution is said to be positively skewed, and vice versa, as in Fig. 103.

If there is only one mode, the median is found empirically to lie between the mode and the mean, the relationship being given approximately by:

$$\textit{Mode} = \text{median} - 3(\text{mean} - \text{median})$$

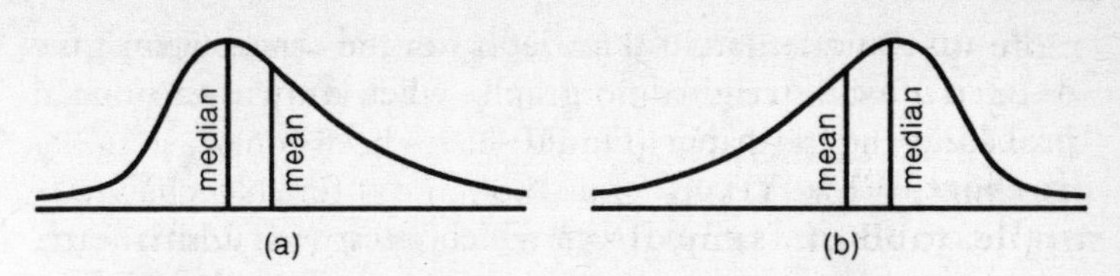

Fig. 103. **Skew.** (a) Positively-skew distribution, where the median < mean. (b) Negatively-skew distribution, where the median > mean.

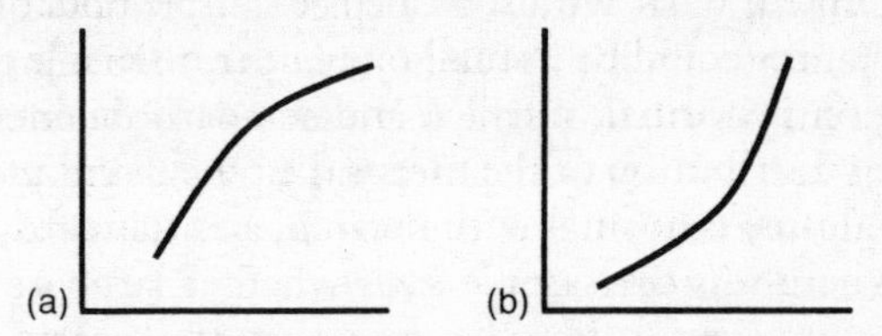

Fig. 104. **Skew.** Lines when distributions are plotted on normal probability graph paper. (a) Positively-skew distribution. (b) Negatively-skew distribution.

Thus, for a unimodal distribution, it is also true that:

$$\text{Mode} < \text{mean} \Rightarrow \text{Positive skew}$$
$$\text{Mode} > \text{mean} \Rightarrow \text{Negative skew.}$$

Skewness can be measured in three ways.

(a) The *coefficient of skewness*, given by:

$$E\left(\frac{X-\mu}{\sigma}\right)^3$$

where μ, σ are the population mean and standard deviation.

(b) The *quartile coefficient of skewness*, given by:

$$\frac{\text{upper quartile} + \text{lower quartile} - 2 \times \text{median}}{\text{upper quartile} - \text{lower quartile}}$$

(c) *Pearson's measure of skewness*, given by:

$$\frac{\text{mean} - \text{mode}}{\text{standard deviation}} \quad \text{or} \quad \frac{3(\text{mean} - \text{median})}{\text{standard deviation}}$$

Positive and negative skew of a normal distribution give distinctive non-straight line graphs when drawn on normal probability graph paper (Fig. 104).

slit chart, see PIE CHART.

small sample, *n.* a sample from which, because of its small size, conclusions have to be treated with particular caution. Hypothesis testing may involve the use of special tests.

A sample is often considered small if it is less than about 30 in size. This does, however, depend on the reason why the sample is being taken; what would be a large sample under one set of circumstances could be a small one under others. If the parent population is normal, mean μ and standard deviation σ, the sampling distribution of the means (i.e., the distribution of the sample means) is normal with mean μ, and standard deviation $\sigma/\sqrt{n}$, where n is the sample size, whether large or small.

It is, however, often the case that the parent standard deviation is unknown and can only be estimated from the sample. If this estimate, $\hat{\sigma}$, is used instead of σ (the true parent standard deviation), the distribution obtained is the t-DISTRIBUTION rather than the NORMAL DISTRIBUTION. Thus, the t-distribution, with n − 1 DEGREES OF FREEDOM, is given by:

$$\frac{\bar{x}-\mu}{\hat{\sigma}/\sqrt{n}}$$

If the sample size is large, the t-distribution is almost the same as the normal distribution, but for small samples the two are markedly different.

Spearman's rank correlation coefficient, see CORRELATION COEFFICIENT.

split-half method, see RELIABILITY COEFFICIENT.

spread, see DISPERSION.

square law graph, see GRAPH.

standard deviation, *n.* a measure of spread or DISPERSION. It is root mean squared deviation. Symbol: σ (parent SD), $\hat{\sigma}$ (estimated parent SD), s (sample SD).

Standard deviation can be calculated from two equivalent formulae.

(a) $\sqrt{\sum_i \frac{x_i^2}{n} - \bar{x}^2}$ (b) $\sqrt{\sum_i \frac{(x_i - \bar{x})^2}{n}}$

For grouped data, these are written:

$\sqrt{\sum_i \frac{f_i x_i^2}{n} - \bar{x}^2}$ $\sqrt{\sum_i \frac{f_i (x_i - \bar{x})^2}{n}}$

For the figures, 12, 9, 9, 9, 8, 6, 6, 5, 4, 2, the two calculations of standard deviation are as follows:

(a)

x	f	fx	fx^2
12	1	12	144
9	3	27	243
8	1	8	64
6	2	12	72
5	1	5	25
4	1	4	16
2	1	2	4
	10	70	568

$n = 10$

Mean $\bar{x} = 7$ *Standard deviation* $= \sqrt{\frac{568}{10} - 7^2} = 2.79$

(b)

x	f	fx	$x - \bar{x}$	$(x - \bar{x})^2$	$f(x - \bar{x})^2$
12	1	12	5	25	25
9	3	27	2	4	12
8	1	8	1	1	1
6	2	12	−1	1	2
5	1	5	−2	4	4
4	1	4	−3	9	9
2	1	2	−5	25	25
	10	70			78

$n = 10$ $\div n = 7.8$

Mean $\bar{x} = 7$ *Standard deviation* $= \sqrt{7.8} = 2.79$

The square of standard deviation is VARIANCE.

For samples of size n drawn from a parent population of size N, with standard deviation σ, the standard deviation of the sample means is given by:

$$\frac{\sigma}{\sqrt{n}}\sqrt{\frac{N-n}{N-1}}$$

which simplifies, for large N, to:

$$\frac{\sigma}{\sqrt{n}}$$

If a sample of size n has standard deviation s, the parent standard deviation σ may be estimated as:

$$\hat{\sigma}=s\sqrt{\frac{n}{n-1}}$$

standard error, *n.* **1.** the standard deviation of the means of samples of given size drawn from a particular parent population; the standard deviation of the sampling distribution of the means. If n is the sample size, N the size of the parent population, and σ the standard deviation of the parent population, standard error is given by:

$$\frac{\sigma}{\sqrt{n}}\sqrt{\frac{N-n}{N-1}}$$

For a large parent population, or for sampling with replacement, this may be simplified to give the commonly-used formula:

$$\frac{\sigma}{\sqrt{n}}$$

Standard error is a measure of a reasonable difference between a sample mean and the parent mean, and is used in tests of whether a particular sample could have been drawn from a given parent population. It is used in working out CONFIDENCE LIMITS and CONFIDENCE INTERVALS.

2. the standard deviation of sample statistics other than the mean.

standard form, *n.* a way of writing numbers which is particularly convenient if they are very large or very small. A number in standard form is of the type, $a \times 10^n$, where a lies between 1 and 10, and n is an integer. The mass of an electron is actually

$$.000\,000\,000\,000\,000\,000\,000\,000\,000\,000\,911\text{kg}$$

This is more conveniently expressed in standard form as 9.11×10^{-31}kg.

standardization, *n.* transformation of the values of the variables of a distribution, so that it has mean 0 and standard deviation 1. Standardization is carried out using the transformation,

$$z = \frac{(x-\mu)}{\sigma}$$

where μ is the mean, σ the standard deviation, x the original value, and z the new value.

Example: a machine makes bolts of mean length 10cm, with standard deviation .05cm. A particular bolt has length 10.12cm. This value is standardized to give:

$$z = \frac{(10.12 - 10.00)}{.05} = 2.4$$

Standardization allows tables for certain distributions to be used, like those for the NORMAL DISTRIBUTION and the T-DISTRIBUTION.

statistic, *n.* any function of sample data, containing no unknown parameters, such as mean, median or standard deviation.

The term statistic is used for a sample, PARAMETER for a parent population. By convention, parameters are often assigned

Greek letters (like μ and σ), statistics Roman letters (e.g. m and s).

stem-and-leaf plot or **stemplot,** *n.* a diagram used to show data graphically, often used in EXPLORATORY DATA ANALYSIS. A stem-and-leaf plot is used to illustrate the major features of the distribution of the data in a convenient and easily drawn form. Example: the stem-and-leaf plot showing the birth rates in Western Europe, per thousand population (1980); Source: *1981 Statistical Yearbook*, UN (1983).

Austria	12.0	Denmark	11.2	Finland	13.1	France	14.8
W. Germany	10.0	Greece	15.9	Iceland	19.8	Ireland	21.9
Italy	11.2	Luxembourg	11.6	Netherlands	12.8	Norway	12.5
Portugal	16.3	Spain	15.1	Sweden	11.7	Switzerland	11.6
UK	13.5	Yugoslavia	17.0			n = 18	

Stem	*Leaves*
9	
10	0
11	2 2 6 6 7
12	0 5 8
13	1 5
14	8
15	1 9
16	3
17	0
18	
19	8
20	
21	9
22	

The figures to the left of the line are called the stem, those to the right the leaves. For example, against the stem figure of 13 are entered the figures for Finland (13.1) and the UK (13.5) in the form of the leaf figures 1 and 5. To make the diagram more compact, OUTLIERS are often listed separately, low values at the low end and high values at the high end.

stemplot, see STEM-AND-LEAF PLOT.

Stirling's approximation, see FACTORIAL.

stochastic, *adj.* (of a variable) exhibiting random behaviour, which cannot be explained fully by a purely deterministic model.

stochastic matrix, see TRANSITION MATRIX.

stochastic process, *n.* a random process, usually a variable measured at a set of points in time or space, which can be in one of a number of states. For example, the length of a queue measured at different times is a stochastic process. In the special case when the next state depends only on the present state, and not on the past history of the process, a stochastic process is called a MARKOV CHAIN.

The probability that a system in a particular state at one time will be in some other state at a later time is given by the *transition probability*.

stratified sampling, *n.* a method of sampling which makes allowance for known differences within the SAMPLE SPACE. For example, an opinion pollster, carrying out a survey of voting intentions before a general election, might decide that those to be interviewed should be drawn from the rural, suburban and urban populations; he would also decide on the proportion of the total sample to be drawn from each of these three strata.

Stratified sampling in general reduces the variability of the estimates produced from the sample. Consequently it is usually preferable to SIMPLE RANDOM SAMPLING.

student's t-test, see t-TEST.

subset, *n.* a set, all of whose members are contained within another set. Thus set A is a subset of set B, if every element in A also belongs to B. This is written $A \subset B$ in set notation.

The set {a,b,c} has the following eight subsets:

$$\{\}\ \{a\}\ \{b\}\ \{c\}\ \{a,b\}\ \{b,c\}\ \{a,c\}\ \{a,b,c\}$$

Notice that the empty set and the set itself are among the subsets.

summation notation, *n.* the use of the symbol Σ to mean 'The sum of ...'. For example, $x_1 + x_2 + \ldots + x_n$ in summation notation is written

$$\sum_{i=1}^{n} x_i$$

This is often written more loosely as,

$$\sum_i x_i \quad \text{or} \quad \sum x$$

meaning the sum of all the values of x.

If x has values 2, 3, 8, 10 and 11, then

$$\sum x = 2 + 3 + 8 + 10 + 11 = 34$$
$$\sum x^2 = 4 + 9 + 64 + 100 + 121 = 298$$

sum of squares, *n.* (usually) the sum of the squares of the deviations from the mean. This can be calculated as:

$$\sum_i (x_i - \bar{x})^2 \quad \text{or} \quad \sum_i x_i^2 - n\bar{x}^2$$

or, for grouped data, (where $n = \sum_i f_i$):

$$\sum_i f_i (x_i - \bar{x})^2 \quad \text{or} \quad \sum_i f_i x_i^2 - n\bar{x}^2$$

Example: calculate the sum of squares for the numbers 4, 4, 5, 8, 9.

	x	f	fx	$x-\bar{x}$	$f(x-\bar{x})^2$
	4	2	8	−2	8
	5	1	5	−1	1
	8	1	8	2	44
	9	1	9	3	9
Σ		5	30		22

$$\textit{Mean}\ \bar{x} = \frac{\sum fx}{\sum f} = \frac{30}{5} = 6$$

Sum of squares = 22

The sum of squares = variance × n, the number of observations.

sum of two or more random variables, *n.* see DIFFERENCE AND SUM OF TWO OR MORE RANDOM VARIABLES.

survey, *n.* an investigation of one or more variables of a population, which may be quantitative or qualitative. Surveys are often, but by no means invariably, carried out on human beings. A survey may involve 100% sampling; for example, a survey of the opinions of all the householders of a village about a proposed by-pass. When this is not the case, the term does, nonetheless, imply adequate sampling. The sampling scheme might then be devised in the light of evidence from a PILOT SURVEY.

systematic error, *n.* error due to the method of collecting or processing data. This may be due to bias in the EXPERIMENTAL DESIGN, or to the ESTIMATOR being used.

systematic sampling, *n.* sampling in which each member of the sample is chosen by ordering the frame in some way, and then selecting individuals at regular intervals. Thus a 1% sample of telephone subscribers could be taken by selecting the 1st, 101st, 201st, 301st (and so on) of the names listed in the telephone directory.

A systematic sample can be given a random start. In the example above, a random number would be chosen between 1 and 100; say 53. Then the subscribers chosen would be those listed at 53rd, 153rd, 253rd, and so on.

A systematic sample has the advantage of being quick and easy to use. If, however, the frame is not suitable, a systematic sample should not be used; this is likely to be the case if there is cyclic variation in the frame.

T

tally, *n.* a method of keeping count in blocks of five.

1	= 1
11	= 2
111	= 3
1111	= 4
~~1111~~	= 5
~~1111~~ 1	= 6
~~1111~~ ~~1111~~ 11	= 12

t-distribution, *n.* the distribution of the means of samples of given size, standardized using estimated values of the parent standard deviation. It is used in the t-TEST.

The distribution of the means of samples of size n, drawn from a normal parent population with mean μ and standard deviation σ, is itself normal with mean μ and standard deviation $\sigma/\sqrt{n}$, whatever the size of the sample. Often, however, the parent standard deviation σ is unknown, and can only be estimated from the sample. If this estimate, $\hat{\sigma}$, is used instead of σ, the distribution obtained for the normalized variable

$$\frac{(\bar{x}-\mu)}{\dfrac{\hat{\sigma}}{\sqrt{n}}}$$

is the t-distribution with $n-1$ degrees of freedom, where $\bar{x}$ is the sample mean.

The t-distribution is different for each value of ν, the number of DEGREES OF FREEDOM. As ν increases, the t-distribution approaches the NORMAL DISTRIBUTION ever more closely, and for $\nu = \infty$ the two distributions are the same (Fig. 105).

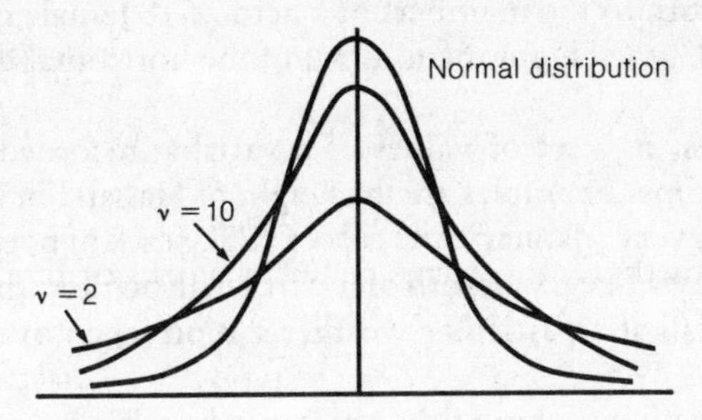

Fig. 105. **t-distribution.** Graph showing t-distributions ($\nu = 2$, and $\nu = 10$) and the normal distribution.

t-distribution tables are usually given only for critical values of t (e.g., 10%, 5%, 2%, 1%, .2%, .1%). To give full tables for a reasonable choice of the number of degrees of freedom ν would be very cumbersome.

test statistic, *n.* **1.** a statistic which is calculated from a sample in order to test a hypothesis about the parent population from which the sample is drawn; examples are χ^2, and t. **2.** a statistic calculated from more than one sample (like F, U, etc.) to compare the parent populations from which they are drawn.

three-quarters high rule, *n.* an EMPIRICAL rule for ensuring that a BAR CHART gives the right impression. It states that the

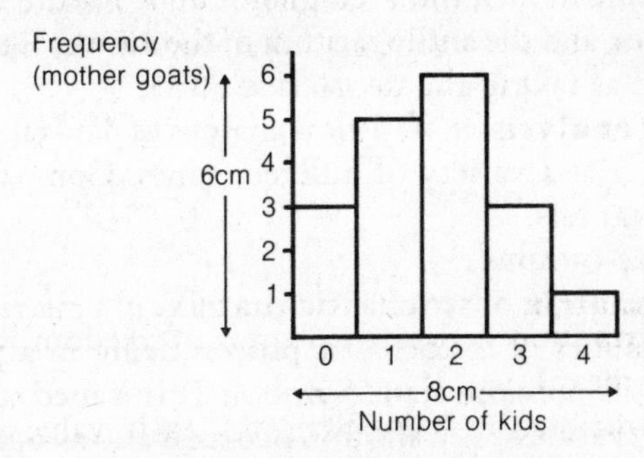

Fig. 106. **Three-quarters high rule.**

maximum height should be three-quarters of the horizontal displacement.

Example: in the case of the bar chart (Fig. 106) of the number of live kids born to the members of a herd of 18 female goats, in spring 1983, the height, 6cm, is 3/4 of the horizontal distance, 8cm.

time series, *n.* a set of values of a variable recorded over a period of time. Examples are the height of Hassan bin Ibrahim measured every 1 January; the cost of a 25kg sack of potatoes on the first day of every month at a particular market; the mean daily rainfall at a particular weather station taken over every month.

The variation within a time series may be regarded as having four components; T TREND; S SEASONAL COMPONENT; C CYCLIC COMPONENT; and V IRREGULAR VARIATION.

In an *additive model*, the value of a particular term is given by:

$$X = X_0 + T + S + C + V$$

where X_0 is a base value.

In a *multiplicative model*, it is given by:

$$X = X_0 \times T \times S \times C \times V$$

The effect of seasonal and cyclic components can be eliminated by taking suitable MOVING AVERAGES. For an additive model, the values of the variable are averaged; for a multiplicative model, the averaging is done for the logarithm of the values, and the antilogarithm of the answer is taken; this is the same as taking the GEOMETRIC MEAN.

time series analysis, *n.* statistical analysis of data taken from a single source at a variety of different times. Compare CROSS-SECTION ANALYSIS.

transect, see QUADRAT.

transition matrix or **stochastic matrix,** *n.* a matrix relating the probabilities of a stochastic process being in a particular state given its previous state. Symbol: T. It is used to model a STOCHASTIC PROCESS. A transition matrix is also associated with every MARKOV CHAIN.

Example: if it is fine on one day, the probability of its being fine on the next day is 2/3; if, however, it is wet, the probability of the next day being fine is 1/2. Thus, if Monday is fine, what is the probability of Wednesday being fine?
The information on probabilities can be set out in table form:

		TODAY'S WEATHER	
		Fine	*Wet*
TOMORROW'S	*Fine*	$\frac{2}{3}$	$\frac{1}{2}$
WEATHER	*Wet*	$\frac{1}{3}$	$\frac{1}{2}$

This can then be written as the transition matrix,

$$\begin{pmatrix} \frac{2}{3} & \frac{1}{2} \\ \frac{1}{3} & \frac{1}{2} \end{pmatrix}$$

The fact that Monday is fine means a probability distribution of:

$$\begin{pmatrix} 1 \\ 0 \end{pmatrix} \begin{matrix} \text{Fine} \\ \text{Wet} \end{matrix}$$

The probability distribution for Tuesday is given by:

Transition matrix	×	Monday's probability distribution	=	Tuesday's probability distribution
$\begin{pmatrix} \frac{2}{3} & \frac{1}{2} \\ \frac{1}{3} & \frac{1}{2} \end{pmatrix}$		$\begin{pmatrix} 1 \\ 0 \end{pmatrix}$		$\begin{pmatrix} \frac{2}{3} \\ \frac{1}{3} \end{pmatrix} \begin{matrix} \text{Fine} \\ \text{Wet} \end{matrix}$

For Wednesday,

$$\begin{pmatrix} \frac{2}{3} & \frac{1}{2} \\ \frac{1}{3} & \frac{1}{2} \end{pmatrix} \begin{pmatrix} \frac{2}{3} \\ \frac{1}{3} \end{pmatrix} = \begin{pmatrix} \frac{11}{18} \\ \frac{7}{18} \end{pmatrix} \begin{matrix} \text{Fine} \\ \text{Wet} \end{matrix}$$

The numbers in each column of a transition matrix must add to 1.

$$\begin{pmatrix} \frac{2}{3} & \frac{1}{2} \\ \frac{1}{3} & \frac{1}{2} \end{pmatrix}$$
$$= \overline{1} \quad \overline{1}$$

Transition matrices need not be 2×2. If instead of defining the weather on any day to be wet or fine, as in the example, three possibilities, fine, mixed or wet weather had been allowed, then a 3×3 transition matrix would have been needed.

Calculations of the PROBABILITY DISTRIBUTION many time intervals later can be simplified by writing the transition matrix T in the form

$$T = QDQ^{-1}$$

where D is the matrix with the Eigen values of T in the leading diagonal and zeros elsewhere, and Q is the matrix consisting of the corresponding Eigen vectors. Thus, the transition matrix in the example above can be written as

$$\begin{array}{ccccccc} T & = & Q & & D & & Q^{-1} \end{array}$$

$$\begin{pmatrix} \frac{2}{3} & \frac{1}{2} \\ \frac{1}{3} & \frac{1}{2} \end{pmatrix} = \begin{pmatrix} 3 & 1 \\ 2 & -1 \end{pmatrix} \begin{pmatrix} 1 & 0 \\ 0 & \frac{1}{6} \end{pmatrix} \begin{pmatrix} \frac{1}{5} & \frac{1}{5} \\ \frac{2}{5} & -\frac{3}{5} \end{pmatrix}$$

where 1 and $\frac{1}{6}$ are the *Eigen values* of T, and

$\begin{pmatrix} 3 \\ 2 \end{pmatrix}$ and $\begin{pmatrix} 1 \\ -1 \end{pmatrix}$ the corresponding *Eigen vectors*.

One of the Eigen values of a transition matrix is always 1.

Writing a transition matrix in this way allows it to be raised to any power, n, using the formula

$$T^n = QD^nQ^{-1}$$

Since the matrix D is in diagonal form, it is easily raised to the power n; for example

$$\begin{pmatrix} 1 & 0 \\ 0 & \frac{1}{6} \end{pmatrix}^n = \begin{pmatrix} 1^n & 0 \\ 0 & (\frac{1}{6})^n \end{pmatrix} = \begin{pmatrix} 1 & 0 \\ 0 & (\frac{1}{6})^n \end{pmatrix}$$

Example: if it is fine one Monday, what is the probability distribution for (a) the following Monday, (b) the same day a year later?

(a) This is given by:

$$\begin{pmatrix} \frac{2}{3} & \frac{1}{2} \\ \frac{1}{3} & \frac{1}{2} \end{pmatrix}^7 \begin{pmatrix} 1 \\ 0 \end{pmatrix}$$

$$= \begin{pmatrix} 3 & 1 \\ 2 & -1 \end{pmatrix} \begin{pmatrix} 1 & 0 \\ 0 & \frac{1}{6} \end{pmatrix}^7 \begin{pmatrix} \frac{1}{5} & \frac{1}{5} \\ \frac{2}{5} & -\frac{3}{5} \end{pmatrix} \begin{pmatrix} 1 \\ 0 \end{pmatrix}$$

$$= \begin{pmatrix} 3 & 1 \\ 2 & -1 \end{pmatrix} \begin{pmatrix} 1 & 0 \\ 0 & 1/279936 \end{pmatrix} \begin{pmatrix} \frac{1}{5} & \frac{1}{5} \\ \frac{2}{5} & -\frac{3}{5} \end{pmatrix} \begin{pmatrix} 1 \\ 0 \end{pmatrix}$$

$$= \begin{pmatrix} .600001 & .599998 \\ .399999 & .400002 \end{pmatrix} \begin{pmatrix} 1 \\ 0 \end{pmatrix}$$

$$= \begin{pmatrix} .600001 \\ .399999 \end{pmatrix} \begin{matrix} \text{Fine} \\ \text{Wet} \end{matrix}$$

(b) The same day next year, 365 days later, the probability distribution is given by:

$$\begin{pmatrix} \frac{2}{3} & \frac{1}{2} \\ \frac{1}{3} & \frac{1}{2} \end{pmatrix}^{365} \begin{pmatrix} 1 \\ 0 \end{pmatrix}$$

$$\begin{pmatrix} \frac{2}{3} & \frac{1}{2} \\ \frac{1}{3} & \frac{1}{2} \end{pmatrix}^{365} \begin{pmatrix} 1 \\ 0 \end{pmatrix} = \begin{pmatrix} 3 & 1 \\ 2 & -1 \end{pmatrix} \begin{pmatrix} 1 & 0 \\ 0 & \frac{1}{6} \end{pmatrix}^{365} \begin{pmatrix} \frac{1}{5} & \frac{1}{5} \\ \frac{2}{5} & -\frac{3}{5} \end{pmatrix} \begin{pmatrix} 1 \\ 0 \end{pmatrix}$$

$$= \begin{pmatrix} 3 & 1 \\ 2 & -1 \end{pmatrix} \begin{pmatrix} 1 & 0 \\ 0 & 0 \end{pmatrix} \begin{pmatrix} \frac{1}{5} & \frac{1}{5} \\ \frac{2}{5} & -\frac{3}{5} \end{pmatrix} \begin{pmatrix} 1 \\ 0 \end{pmatrix}$$

$$= \begin{pmatrix} \frac{3}{5} \\ \frac{2}{5} \end{pmatrix} \begin{matrix} \text{Fine} \\ \text{Wet} \end{matrix}$$

(This case is, in effect, one where $n = \infty$. The result in such a case gives the unconditional probability distribution.) Thus, in this example, the probability of a day one year hence being fine is 3/5, wet 2/5.

transition probability, see STOCHASTIC PROCESS.

tree diagram, *n.* a diagram in which events or relationships are represented by points joined by lines. It is laid out so that the rules, 'multiply along the branches', and 'add vertically' apply.

Example: a man is lost in the jungle. He comes to a fork in the path. He does not know it, but if he turns left, he has to pass a man-eating tiger who sleeps 8 hours a day, but will catch and eat him if awake. If he turns right he passes a cannibal's hut. The cannibal spends 3 days a week away; if he is at home he will certainly make a meal of the man. What is the probability that the man gets eaten if he is equally likely to turn left or right?

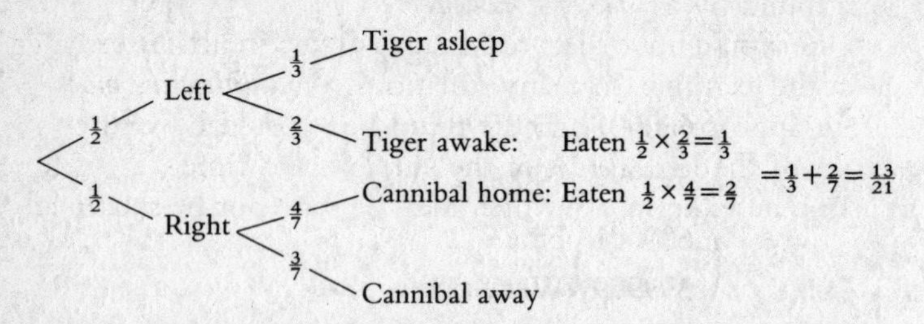

He therefore has a probability of 13/21 of being eaten.

trend, *n.* the underlying direction and rate of change in a TIME SERIES, when allowance has been made for the effects of SEASONAL COMPONENTS, CYCLIC COMPONENTS, and IRREGULAR VARIATION. The equation of a trend may be called the *trend line*; it need not be straight although it often is (Fig. 107).

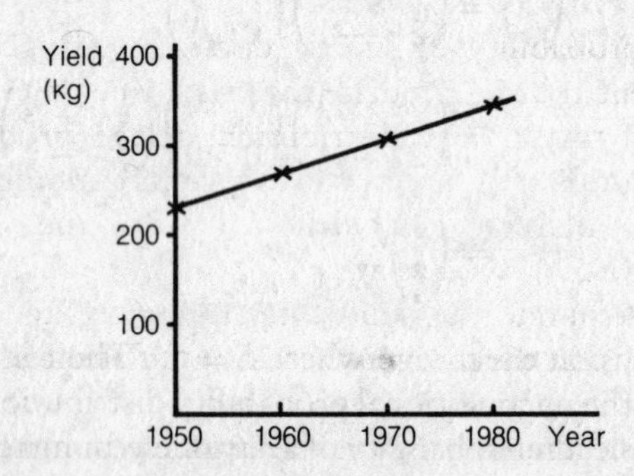

Fig. 107. **Trend.** The straight trend line for production of truffles from 1950 to 1980.

Example: the UK truffle production from 1950 to 1980 was as follows:

Year	1950	1960	1970	1980
Truffle yield (kg)	230	270	310	350

It is easy to see that these figures fit the trend line,

$$\text{Yield} = 230 + 4t$$

where t is the number of years since 1950. In this case, the yield was found by an *additive model*.

One would not expect real figures to fit a trend line exactly, as in the example. In many situations, a *multiplicative model* is more appropriate, when the trend line would be written in terms of the logarithms of the variable and time.

trial, *n.* an experiment which may or may not be successful. Examples:
Tossing a coin: success is getting a head.
Throwing a die: success is getting a 6.
Selecting a housewife at random: success is if she uses a particular brand of soap.
Selecting a pepper moth: success is if it is the dark variety.

It is essential that success be clearly defined so that a trial is either successful or unsuccessful. Trials are said to be independent if the probability of success in one is not influenced by the outcome of any of the others.

If the probability of success does not change from one independent trial to another, the trials are sometimes called BERNOULLI TRIALS. The distribution of the probabilities of different numbers of successes (0 to n) in n Bernoulli trials is the BINOMIAL DISTRIBUTION, also called the BERNOULLI DISTRIBUTION.

The GEOMETRIC DISTRIBUTION, for $n=1$ to ∞, is the probability that the first success will occur at the nth Bernoulli trial.

PASCAL'S DISTRIBUTION, (or the negative binomial distribution) gives the probability that the rth success occurs at the nth Bernoulli trial, where n takes values from r to infinity.

t-test or **students' t-test,** *n.* a test used on the means of small samples of one of these NULL HYPOTHESES:

(a) That the sample has been drawn from a population of

given mean and unknown standard deviation (which has therefore to be estimated from the sample).

(b) That two samples have both been drawn from the same parent population.

Example (one sample): the annual rainfall at a weather station has been found over many years to have a mean value of 120cm. The rainfall for 10 years, 1975–84, has been 130, 151, 109, 138, 156, 145, 101, 129, 138 and 103cm. Is there evidence, at the 5% SIGNIFICANCE LEVEL, of any change in annual rainfall?

Null hypothesis: The mean rainfall is unaltered; parent mean $\mu = 120$.

Alternative hypothesis: The mean rainfall has changed (a two-tail test). See ONE- AND TWO-TAIL TESTS.

Since the sample is small, and the parent standard deviation is unknown, a t-test should be used.

The null hypothesis $\Rightarrow$ parent mean, $\mu = 120$.

From the sample figures, we calculate

Sample size $n = 10$
Sample mean $\bar{x} = 130$
Sample SD $s = 18.66$

Estimated parent SD $\hat{\sigma} = s\sqrt{\dfrac{n}{n-1}} = 19.67$

The STATISTIC t is given by:

$$t = \frac{\bar{x} - \mu}{\hat{\sigma}/\sqrt{n}} = \frac{130 - 120}{19.67/\sqrt{10}} = 1.608$$

There are $10 - 1 = 9$ DEGREES OF FREEDOM. For 9 degrees of freedom (see table p. 248) the critical value for t at the 5% level is 2.262. Since $1.608 < 2.262$, the null hypothesis is accepted at the 5% significance level; the evidence does not support the claim that the mean rainfall has changed.

Example (two samples): a geologist has two samples of ore, A and B. Sample A has 7 specimens; the mass of metal per kilogram of ore in them has mean value 81g, with standard deviation 8g. Sample B, with 10 specimens, has mean value

72g of metal per kg of ore, with standard deviation 6g. He wishes to test, at the 5% significance level, whether the two samples have been drawn from the same source.
Null hypothesis: The two samples are taken from the same parent population.
Alternative hypothesis: They are drawn from parent populations with different means (a two-tail test).
If the null hypothesis is true, the distribution of $(m_1 - m_2)$, where m_1 and m_2 are the two sample means, should have:

$$\textit{Mean} = 0$$

$$\textit{Standard deviation} = \sqrt{\frac{\sigma^2}{n_1} + \frac{\sigma^2}{n_2}}$$

where n_1 and n_2 are the sizes of the two samples.
The best ESTIMATOR for σ, the parent standard deviation, from two samples is:

$$\sqrt{\frac{(n_1 s_1^2 + n_2 s_2^2)}{(n_1 + n_2 - 2)}}$$

The degrees of freedom ν are $n_1 + n_2 - 2$.
In this case the estimated parent standard deviation is:

$$\sqrt{\frac{(7 \times 8^2 + 10 \times 6^2)}{(7 + 10 - 2)}} = 7.34$$

and so the standard deviation of the distribution of $m_1 - m_2$ is

$$\sqrt{\left(\frac{7.34^2}{7} + \frac{7.34^2}{10}\right)} = 3.62$$

The difference in the two sample means is $81 - 72 = 9$, so

$$t = \frac{9 - 0}{3.62} = 2.49$$

The critical value of t for $\nu = 15$ $(= 7 + 10 - 2)$ at the 5% significance level is 2.13. Since $2.49 > 2.13$ the null hypothesis is rejected at this level; the geologist would not be justified in thinking the two samples came from the same source.

The t-test assumes that the variances of the two populations are equal. In this example, that would be the case if the two samples of ore were drawn from the same source.

PAIRED SAMPLES or MATCHED SAMPLES are frequently small in size and so the t-test is often used on them. Compare F-TEST.

two-sample test, *n.* a test on two different, non-matched samples. Compare PAIRED-SAMPLE TEST. This term is often used to mean a test of the NULL HYPOTHESIS that two samples are drawn from parent populations with the same mean, or the same parent population. It is this test which is the subject of the rest of this entry.

The notation used is that the samples have sizes n_1, n_2, means m_1, m_2, and standard deviations s_1, s_2. Their parent populations, if different, have standard deviations σ_1, σ_2; if they are taken to be drawn from the same parent population, this has standard deviation σ.

The null hypothesis may then be stated as follows: The sampling distribution of the differences of the means $m_1 - m_2$ has mean $= 0$.

This distribution has standard deviation

$$\sqrt{\frac{\sigma^2}{n_1} + \frac{\sigma^2}{n_2}} \quad \text{or} \quad \sqrt{\frac{\sigma_1^2}{n_1} + \frac{\sigma_2^2}{n_2}}$$

according to whether the parent population is common to both samples or not. It is, however, likely that the parent standard deviation is not known. In that case it must be estimated from the sample data. If it is reasonable to assume a common parent population, (a necessary assumption if the t-TEST is to be used), its standard deviation $\hat{\sigma}$ may be estimated as:

$$\hat{\sigma} = \sqrt{\frac{(n_1 s_1^2 + n_2 s_2^2)}{(n_1 + n_2 - 2)}}$$

If the samples are small and the parent standard deviation has been estimated, a t-test is applied on the sample data. If the samples are not small, or the parent standard deviation(s) is

(are) known, then the NORMAL DISTRIBUTION can be used.
Example: a sample of 100 harvest mice collected in England had mean mass 21.52g, standard deviation 4.81g. A second sample of 150 harvest mice was collected in Scotland: they had mean mass 19.42g, with standard deviation 4.60g. Is there evidence, at the 5% SIGNIFICANCE LEVEL, to suggest that harvest mice in England and Scotland have different mean masses?
Null hypothesis: There is no difference in their mean masses; both samples are drawn from parent populations with the same mean.
Alternative hypothesis: There is a difference in the mean masses of harvest mice in England and Scotland.
Test: Two-tail test using the normal distribution (samples are not small).
Significance level: 5%.
CRITICAL REGION: $|z| > 1.96$.
Observed difference in means $= 21.52 - 19.42 = 2.10$g.
Estimated standard deviation of sampling distribution is

$$\sqrt{\frac{s_1^2}{n_1} + \frac{s_2^2}{n_2}} = \sqrt{\left(\frac{4.81^2}{100} + \frac{4.60^2}{150}\right)} = 0.610$$

Mean of sampling distribution of $(m_1 - m_2) = 0$ (see Fig. 108).

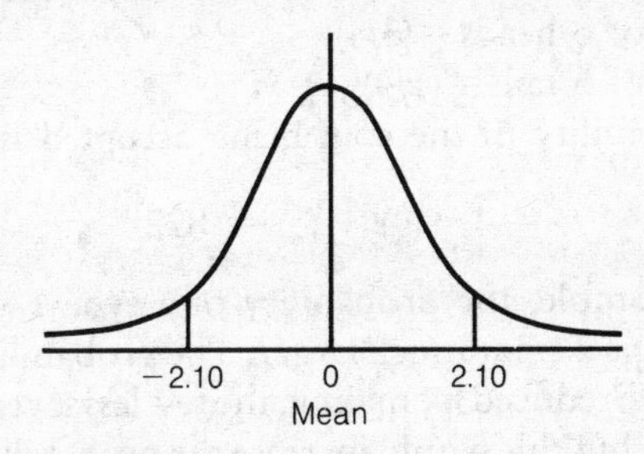

Fig. 108. **Two-sample test.** Sampling distribution of $(m_1 - m_2)$. SD = 0.610.

So the z-value is:

$$\frac{2.10 - 0}{0.610} = 3.44$$

Since $3.44 > 1.96$ the null hypothesis is rejected. It is concluded that there is a difference in the mean masses of harvest mice in England and Scotland.

two-tail test, see ONE- AND TWO-TAIL TESTS.

two-way analysis of variance, see ANALYSIS OF VARIANCE, FRIEDMAN'S TWO-WAY ANALYSIS OF VARIANCE BY RANK.

type 1 and type 2 errors, *n.* those errors which occur when, respectively, a true NULL HYPOTHESIS is rejected and a false null hypothesis is accepted.

Example: coins to be used in a casino are tested for bias by being tossed six times. If a coin comes either heads or tails all six times it is rejected.

(a) What is the probability that a good coin is rejected (type 1 error)?

Null hypothesis: The coin is not biased.

The probability of a head = probability of a tail = 1/2; thus the probability of 6 heads is $(1/2)^6$, of 6 tails is $(1/2)^6$. So the probability of a good coin being rejected is:

$$(\tfrac{1}{2})^6 + (\tfrac{1}{2})^6 = \tfrac{1}{32}$$

(b) A coin is in fact biased with probability 1/4 of coming tails. What is the probability that it is accepted (type 2 error)?

Probability of 6 heads = $(3/4)^6$

Probability of 6 tails = $(1/4)^6$

So the probability of the coin being accepted is:

$$1 - (\tfrac{3}{4})^6 - (\tfrac{1}{4})^6 = .822$$

In this example, the probability of a type 1 error is quite small, of a type 2 error much larger. The probability of a type 2 error could be reduced by making the test less severe (i.e., fewer than 6 losses) but this would increase the probability of a type 1 error. In this particular case, a type 1 error (rejecting a good coin) would not matter much, but a type 2 error (accepting a biased coin) would be serious. Consequently, the test being applied is quite unsuitable.

A better test would be to toss the coin 6 times, and accept it only if the results were 3 heads and 3 tails. In that case, the probability of a type 1 error, rejecting the unbiased coin, is .6875. The probability of a type 2 error, accepting a biased coin, is .132 in the case when the probability of it coming tails is 1/4. A more severe test (e.g. tossing coins 20 times and accepting only those that come 10 heads and 10 tails) would result in a higher number of unbiased coins being rejected, but fewer biased ones being accepted.

Typical Year Index, see INDEX NUMBER.

U

unbiased estimator, see BIASED ESTIMATOR.

uniform distribution, see RECTANGULAR DISTRIBUTION.

union, *n.* a SET containing all members of two given sets. The union of sets A and B is written in set notation as $A \cup B$. For example, if $P = \{p,e,t,s\}$ and $C = \{c,a,t,s\}$, then $P \cup C = \{p,e,t,s,c,a\}$.

universal set, *n.* the SET of all elements from which the members of a particular set can be drawn. Symbol: $\mathscr{E}$. For example, if $A = \{a,c,g,s\}$, $\mathscr{E}$ could be the set of all the letters in the alphabet; if S = {Penny Red stamps}, $\mathscr{E}$ might be {All postage stamps} or {British postage stamps}.

V

variable error, *n.* error occurring in certain types of experiments on living subjects due to factors which vary during the course of an experiment, such as tiredness, or experience.

variable testing, *n.* testing a product for quality control when the result is a quantitive measurement like the breaking strain of a fishing line. Variable testing may be contrasted with ATTRIBUTE TESTING where the result is a verdict of either 'good' or 'defective' as, for example, in a safety check.

variance, *n.* a measure of DISPERSION, the square of standard deviation.

Variance is calculated from:

Sample data: symbol s^2, s_x^2
Ungrouped (n values)

$$\sum_{i=1}^{n} \frac{(x_i - \bar{x})^2}{n} \qquad \text{or} \qquad \sum_{i=1}^{n} \frac{x_i^2}{n} - \bar{x}^2$$

Grouped (k groups with frequencies $f_1, \ldots, f_k$)

$$\sum_{i=1}^{k} \frac{(f_i[x_i - \bar{x}]^2)}{n} \qquad \text{or} \qquad \sum_{i=1}^{k} \frac{f_i x_i^2}{n} - \bar{x}^2$$

Theoretical distributions: symbol Var (X)

CONTINUOUS VARIABLE $\int x^2 f(x)\,dx - [E(X)]^2$ (probability density function $f(x)$)

DISCRETE VARIABLE $\sum_{i=1}^{n} x_i^2 p(x_i) - [E(X)]^2$ (probability distribution $p(x_i)$ where $1 \leqslant i \leqslant n$ n is a positive integer and may be infinity)

For variables X and Y, and constants a and b,

$$\mathrm{Var}(aX) = a^2\,\mathrm{Var}(X)$$
$$\mathrm{Var}(X + b) = \mathrm{Var}(X)$$
$$\mathrm{Var}(X \pm Y) = \mathrm{Var}(X) + \mathrm{Var}(Y) \pm 2\,\mathrm{cov}(X, Y).$$

Variance and standard deviation are used extensively in statistics. Although it is not easy to give a conceptual meaning to variance, it is important because it occurs frequently in theoretical work. The F-TEST is carried out on the variances of two samples, or one sample and a possible parent population. The KRUSKAL-WALLIS ONE-WAY ANALYSIS OF VARIANCE and FRIEDMAN'S TWO-WAY ANALYSIS OF VARIANCE BY RANK are carried out on the variances of the ranks of samples.

variate, *n.* **1.** a random variable. **2.** a numerical value taken by a random variable.

variation, *n.* dispersion or spread about some value. For example, variation about the mean is called DEVIATION.

Venn diagram or **Euler diagram,** *n.* a diagram in which SETS are represented by areas whose relative positions give the relationships between the sets. Individual elements of the sets may (or may not) be marked in. See Fig. 109.

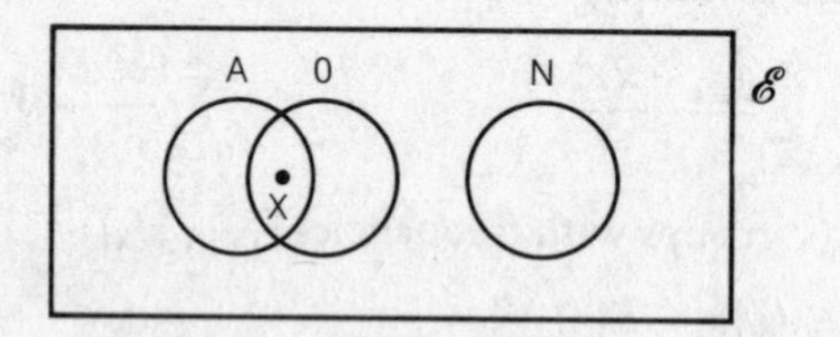

Fig. 109. **Venn diagram** or **Euler diagram.** ℰ = {countries of the world}, A = {African countries}, O = {OPEC countries}, N = {NATO member countries} and X = Nigeria. Thus, it can be seen that Nigeria is both an African and an OPEC country, and no NATO country is African or in OPEC.

vertical line chart, *n.* a method of displaying information similar to a bar chart but using lines instead of bars (see Fig. 110). Compare BAR CHART.

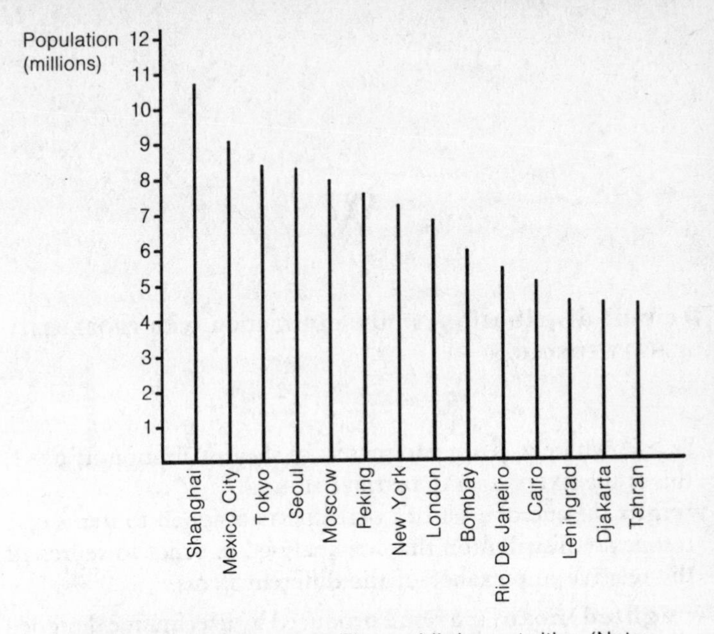

Fig. 110. **Vertical line chart.** The world's largest cities. (Note; figures for city sizes vary considerably with the source; this is due in part to different definitions of city boundaries.) Source: *UN Demographic Year Book*, 1980.

W

Weibull distribution, *n.* the distribution with PROBABILITY DENSITY FUNCTION:

$$f(x) = \alpha\beta x^{\beta-1} e^{-\alpha x^{\beta}}$$

$(x > 0)$ where α, β are parameters of the distribution. If $\beta = 1$, this is the EXPONENTIAL DISTRIBUTION.

weight, *n.* one of a set of COEFFICIENTS assigned to items of a frequency distribution that are analysed in order to represent the relative importance of the different items.

weighted mean, *n.* a result produced by a technique designed to give recognition to the importance of certain factors when compiling the average of a group of values. In general the weighted mean is:

$$\frac{w_1x_1 + w_2x_2 + \ldots + w_nx_n}{w_1 + w_2 + \ldots + w_n}$$

where $w_1, w_2, \ldots, w_n$ are the weightings given to the variables $x_1, x_2, \ldots, x_n$.

Example: a school examines its pupils in Mathematics, English, Science and Humanities. These subjects are given weights, 2 : 3 : 2 : 1. What is the weighted mean mark of Abdul Qadir, whose marks are Mathematics 66%, English 52%, Science 71% and Humanities 42%?

The weighted mean is:

$$\frac{(2 \times 66) + (3 \times 52) + (2 \times 71) + (1 \times 42)}{2 + 3 + 2 + 1}$$

$$= 59\%$$

Wilcoxon matched-pairs signed-rank test, *n.* a non-parametric test of the NULL HYPOTHESIS that two MATCHED SAMPLES are drawn from the same parent population. The technique used in this test is illustrated by the following example.

Example: ten army recruits were given a shooting test. They were then given a lecture on good shooting technique, at the end of which they repeated the test. The results of the two tests, with part of the calculation, are in the table below. Can it be claimed that their performance was better, at the 5% SIGNIFICANCE LEVEL, after the lecture?

Null hypothesis: The lecture had no effect. Any differences between the two sets of scores are due to chance only.

Alternative hypothesis: Their performance improved after the lecture (a one-tail test), see ONE- AND TWO-TAIL TESTS.

Recruit	*First shoot*	*Second shoot*	*Difference*	*Signed rank of difference*	*Ranks with smaller sum*
A	45	46	+1	$+1\frac{1}{2}$	
B	32	30	−2	−3	−3
C	24	30	+6	+6	
D	12	12	0	—	
E	15	19	+4	+5	
F	41	40	−1	$-1\frac{1}{2}$	$-1\frac{1}{2}$
G	28	35	+7	+7	
H	29	40	+11	+9	
I	31	39	+8	+8	
J	25	22	−3	−4	−4
				Total T	$-8\frac{1}{2}$

The *Difference* column is the improvement between the first shoot and the second. Having established this, the absolute values of the numbers in the difference column are then ranked, the largest being given the highest rank. All differences of 0 are ignored. This reduces the number of pairs under consideration from 10 to 9, since the D pair has difference 0. A + sign indicates the difference was positive (i.e., an improvement in shooting), a − sign that it was negative (a worse performance). It is important to realise that the ranking is not done largest to smallest: in that case, −3 would be the lowest rank. It is the absolute values that are ranked, so that values of +1 and −1 would tie for the lowest place.

The right-hand column contains only the negative entries of the ranks; these are summed to give the total T, in this case $-8\frac{1}{2}$. This value (ignoring the − sign) is compared to the critical value found in the tables on page 262 for N=9, significance level .05, for a one-tail test. The critical value for T is found to be 8.

If the total found for T is less than the critical value, then the null hypothesis is rejected. In this case 8.5>8, so the null hypothesis cannot, at the 5% level, be rejected, and the lecture cannot be claimed to have done any good.

If N is not small, T is approximately normally distributed when the null hypothesis is true, with:

$$Mean = \frac{N(N+1)}{4}$$

$$Variance = \frac{N(N+1)(2N+1)}{24}$$

Y

Yates' correction, *n.* a CONTINUITY CORRECTION which may be applied when using the CHI-SQUARED TEST on data from a discrete distribution. It is usually used only in tests on 2×2 contingency tables with small frequencies. It is applied by subtracting 0.5 from the values of O which are greater than E and adding 0.5 to the values of O which are less than E. (O and E are the Observed and Expected frequency respectively in the four categories).

Example: a small colony of a previously unknown species of rat is discovered on a remote island. Some of them have spotted fur on their stomachs, others do not; the numbers are shown in the contingency table.

O		*Male*	*Female*	*Total*
	Spots	6	14	20
	No spots	11	9	20
	Total	17	23	40

Do the figures give reason, at the 5% SIGNIFICANCE LEVEL, to think that the females are more likely to have spots?

NULL HYPOTHESIS: The proportions of males and females having spots are the same.

On this hypothesis the expected values are given in the following table:

E		*Male*	*Female*	*Total*
	Spots	8.5	11.5	20
	No spots	8.5	11.5	20
	Total	17	23	40

8.5, 11.5 are in the ratio 17:23, and sum to 20.

Using Yates' correction,

$$\chi^2 = \frac{(6-8.5+.5)^2}{8.5} + \frac{(14-11.5-.5)^2}{11.5} + \frac{(11-8.5-.5)^2}{8.5} + \frac{(9-11.5+.5)^2}{11.5} = 1.64$$

The number of DEGREES OF FREEDOM, $\nu = (2-1) \times (2-1) = 1$. For $\nu = 1$, at the 5% significance level, the CRITICAL VALUE for $\chi^2 = 3.84$ (see table, p. 000). Since $1.64 < 3.84$ the null hypothesis is accepted. The evidence does not support the conclusion that females are likely to have more spots than males, at the 5% level.

Had Yates' correction not been applied, the value of χ^2 would have been 2.56 rather than 1.64, but that would not, in this case, have altered the conclusion reached.

Yule's coefficient of association, see ASSOCIATION.

Z

z-transformation, see FISHER'S Z-TRANSFORMATION.

z-value, *n.* the standardized value of a random variable. Standardization is carried out by subtracting the mean, and dividing by the standard deviation. Thus the z-value is given by:

$$z = \frac{(x - \mu)}{\sigma}$$

The z-value is needed when using standardized tables such as, for example, those for the NORMAL DISTRIBUTION.

APPENDIX A: List of symbols

A_1, A_2, A_3	acceptance numbers
a	assumed mean
$B(a, b)$	beta function
B(n, p)	binomial distribution
C	cyclic variation
c	class interval
cov	population covariance
${}^nC_r$	binomial coefficient
$d_1, d_2, \ldots$	residuals
E	expected frequencies (χ^2 test)
E( )	expectation
e	base of natural logarithms 2.71 828 ...
e^x, $\exp(x)$	exponential of x
F	distribution, statistic or test
f	frequency
f( )	function, probability density function
F(X)	distribution function
G( )	probability generating function
H	statistic used in Kruskal-Wallis one-way analysis of variance
H_0	null hypothesis
H_1	alternative hypothesis

$\ln$, $\log_e$ natural log
$\log_{10}$ logarithm to base 10

M statistic used in Friedman's two-way analysis of variance
m sample mean
$M_X(t)$ moment generating function of X

N number of groups (χ^2 test)
N population size
N sample size (quality control)
n sample size
n acceptance number
$N(\mu, \sigma^2)$ normal distribution

O observed frequencies (χ^2 test)

p probability
p proportion of defectives
p proportion
$\hat{p}$ estimated probability
p_0, p_n base and current year prices
$P(\mu)$ Poisson distribution
${}^nP_r$ permutation of n objects selected r at a time

q ($= 1 - p$) the probability of failure in a trial
q_0, q_n, q_t base, current and typical year weightings
Q_1, Q_2, Q_3 upper quartile, median, lower quartile

r Pearson's product-moment correlation coefficient
R_1, R_2 sums of ranks (Mann Whitney U-test)

S seasonal component
S sum of squares
s, s_x, $s(x)$ sample standard deviation

s^2, s_x^2, s_y^2	sample variance
s_{xy}	sample covariance
SD	standard deviation
T	statistic used in Wilcoxan matched-pairs signed-rank test
T	trend
T	transition matrix
t	statistic used in t-test
t	t-distribution
t_i	number of ties at rank *i* in Mann Whitney U-test
T_i	correction for ties in Mann Whitney U-test
u	number of rows in a contingency table
U, U_1, U_2	statistics used in Mann Whitney U-test
V	irregular variation
v	number of columns in a contingency table
Var, V, v, $v(x)$	variance
W	Kendall's coefficient of concordance
$w_1, w_2 \ldots$	weightings
X	variable name
x, x_i, x_r	variable value
x	horizontal axis of Cartesian graph
Y	variable name
y, y_i, y_r	variable value
y	vertical axis of a Cartesian graph
z	standardized value
z-	referring to Fisher's z-transformation

α, β	alternatives for 1, 2 in type 1 and 2 errors
$\Gamma(\,)$	gamma function
μ	parent population mean
$\mu_0, \mu_1, \mu_2 \ldots$	0th, 1st, 2nd ... moments
ν	degrees of freedom
π	pi = 3.14157265 ...
ρ	Spearman's rank correlation coefficient
ρ, ρ_0	parent population correlation coefficient
Σ	summation
σ	parent population standard deviation
$\hat{\sigma}$	estimated parent population standard deviation
τ	Kendall's rank correlation coefficient
$\Phi(x)$	distribution function of normal distribution
$\varphi, \varphi(x)$	probability density function of normal distribution
χ^2	distribution, statistic and test

$\lvert\ \rvert$	absolute value, modulus
$\vert$	given
!	factorial
$\in$	is an element of ... (sets)
$\mathscr{E}$	universal set
$\emptyset$	empty set
$\{\,\}$	set
$\cap$	intersection of sets
$\cup$	union of sets
$\subset$	is a subset of
$\{\,\}'$	complement of a set
$\int$	integral
$\bar{x}$	mean value of x
$\hat{}$	estimated

APPENDIX B: List of formulae

Probability Symbol: p

Addition law $p(A \cup B) = p(A) + p(B) - p(A \cap B)$

Multiplication law $p(A \cap B) = p(A|B) \, . \, p(B)$

Bayes' theorem $$p(A|B) = \frac{p(B|A)p(A)}{p(B|A)p(A) + p(B|A')p(A')}$$

Probability distribution Symbol: $p(x)$

$$\sum_i p(x_i) = 1$$

$$\textit{Mean} = \sum_i x_i p(x_i)$$

$$\textit{Variance} = \sum_i x_i^2 p(x_i) - \text{Mean}^2$$

Probability density function (pdf) Symbol: $f(x)$

$$\int f(x) dx = 1$$

$$\textit{Mean} = \int x f(x) dx$$

$$\textit{Variance} = \int x^2 f(x) dx - \text{Mean}^2$$

Probability generating function (pgf) Symbol: G

$$G(1) = 1$$

$$G'(1) = E(X)$$

$$\text{Var}(X) = G''(1) + G'(1) - [G'(1)]^2$$

Mean

$$\textit{Arithmetic mean} = \frac{x_1 + x_2 + \ldots x_n}{n}$$

$$\textit{Geometric mean} = \sqrt[n]{x_1 x_2 \ldots x_n}$$

$$\textit{Harmonic mean} = \frac{1}{\frac{1}{n}\left\{\frac{1}{x_1} + \frac{1}{x_2} + \ldots \frac{1}{x_n}\right\}}$$

$$\textit{Weighted mean} = \frac{w_1 x_1 + w_2 x_2 + \ldots + w_n x_n}{w_1 + w_2 + \ldots w_n}$$

Standard deviation ($\sqrt{\text{Variance}}$)

$$= \sqrt{\left\{\sum_i \frac{x_i^2}{n} - \bar{x}^2\right\}} \quad \text{or} \quad \sqrt{\left\{\sum_i \frac{(x_i - \bar{x})^2}{n}\right\}}$$

For grouped data

$$\sqrt{\left\{\sum_i \frac{f_i x_i^2}{n} - \bar{x}^2\right\}} \quad \text{or} \quad \sqrt{\left\{\sum_i \frac{f_i (x_i - \bar{x})^2}{n}\right\}}$$

Sheppard's correction

$$\text{Corrected Variance} = \text{calculated Variance} - \frac{c^2}{12}$$

Charlier's check

$$\sum_i f_i (x_i + 1)^2 = \sum_i f_i x_i^2 + 2\sum_i f_i x_i + \sum_i f_i$$

Sample covariance

$$s_{xy} = \sum_i \frac{f_i x_i y_i}{n} - \bar{x}\bar{y}$$

$$\text{or} \quad \frac{1}{n}\sum_i f_i (x_i - \bar{x})(y_i - \bar{y})$$

Expectation and variance

$$E(X_1 \pm X_2) = E(X_1) \pm E(X_2) : E(\lambda X) = \lambda E(X)$$

$$E(X^2) = [E(X)]^2 + V(X)$$

$$E(X_1 X_2) = E(X_1)E(X_2) + \text{cov}(X_1, X_2)$$

$$\text{Var}(X_1 \pm X_2) = \text{Var}(X_1) + \text{Var}(X_2) \pm 2\,\text{cov}(X_1, X_2)$$

Regression lines

y on x $\qquad y - \bar{y} = (x - \bar{x}) \dfrac{s_{xy}}{s_x^2}$

x on y $\qquad x - \bar{x} = (y - \bar{y}) \dfrac{s_{xy}}{s_y^2}$

Correlation coefficients

Spearman's $\rho = 1 - \dfrac{6\Sigma D^2}{n(n^2 - 1)}$

Kendall's $\tau = \dfrac{S}{\frac{1}{2}n(n - 1)}$

Pearson's product moment $r = \dfrac{s_{xy}}{s_x \cdot s_y}$

Fisher's z-transformation (normally distributed)

$$z = \tfrac{1}{2} \ln \frac{(1 + r)}{(1 - r)} \text{ is } N\left(z_0, \frac{1}{n-3}\right)$$

The inverse transformation is $r = \text{invtanh } z$

Quantiles

Median: ranked at $\frac{1}{2}(n + 1)$

Upper quartile: ranked at $\frac{3}{4}(n + 1)$

Lower quartile: ranked at $\frac{1}{4}(n + 1)$

Inter-quartile range = Upper quartile–Lower quartile

Distributions

Binomial distribution. Symbol: B(n, p)

$$\text{pgf} = (q + pt)^n = \sum_{r=0}^{n} {}^nC_r q^{n-r}(pt)^r$$

$${}^nC_r = \frac{n!}{r!(n-r)!}$$

Mean = np

Standard deviation = $\sqrt{npq}$

Normal distribution. Symbol: N(μ, σ^2)

$$\text{pdf} = \frac{1}{\sigma\sqrt{2\pi}} e^{-\frac{1}{2}\left(\frac{x-\mu}{\sigma}\right)^2} \quad \text{p}^{\text{df}}$$

Mean = μ
Standard deviation = σ

Poisson distribution. Symbol: P(μ)

$$\text{pgf} = e^{-\mu} \sum_{r=0}^{\infty} \frac{(\mu t)^r}{r!}$$

Mean = μ
Variance = μ

Sampling

Greek letters for parent population parameters, Roman for sample statistics.

Estimators

Single sample:

Parent mean $\hat{\mu}$ = Sample mean m

$$\textit{Parent SD } \hat{\sigma} = s\sqrt{\frac{n}{n-1}}$$

Two samples:

$$\textit{Parent mean } \hat{\mu} = \frac{m_1 n_1 + m_2 n_2}{n_1 + n_2}$$

$$\textit{Parent SD } \hat{\sigma} = \sqrt{\frac{n_1 s_1^2 + n_2 s_2^2}{n_1 + n_2 - 2}}$$

k samples:

$$\textit{Parent mean } \hat{\mu} = \frac{m_1 n_1 + m_2 n_2 \ldots + m_k n_k}{n_1 + n_2 \ldots + n_k}$$

$$\textit{Parent SD } \hat{\sigma} = \sqrt{\frac{n_1 s_1^2 + n_2 s_2^2 \ldots + n_k s_k^2}{n_1 + n_2 \ldots + n_k - k}}$$

Sampling distributions

Of means: normal (or approximately so)

$$\textit{Mean} = \mu$$

$$SD = \frac{\sigma}{\sqrt{n}} \sqrt{\frac{N - n}{N - 1}}$$

or, for large N, or sampling with replacement,

$$SD = \frac{\sigma}{\sqrt{n}}$$

Of difference in means of two samples from different parent populations:

$$\textit{Mean} = \mu_1 - \mu_2$$

$$SD = \sqrt{\frac{\sigma_1^2}{n_1} + \frac{\sigma_2^2}{n_2}}$$

Of difference in means of two samples from the same parent population:

$$\textit{Mean} = 0$$

$$SD = \sqrt{\frac{\sigma^2}{n_1} + \frac{\sigma^2}{n_2}}$$

Quality Control

Average outgoing quality (AOQ)

$$P(p)\left(1 - \frac{n}{N}\right)p$$

Tests

Chi-squared χ^2-test:

$$\chi^2 = \sum \frac{(0-E)^2}{E}$$

Kruskal-Wallis one-way analysis of variance H:

$$H = \frac{12}{N(N+1)}\left(\sum_{j=1}^{k} \frac{R_j^2}{N_j}\right)$$

Friedman's two-way analysis of variance by rank M:

$$M = \frac{12}{NK(K+1)}\left(\sum_{i=1}^{K} R_i^2\right) - 3N(K+1)$$

Mann Whitney U-test:

$$U_1 = n_1 n_2 + \frac{n_1(n_1+1)}{2} - R_1$$

$$U_2 = n_1 n_2 + \frac{n_2(n_2+1)}{2} - R_2$$

Index numbers,

where p_0 = Base year price, p_n = Current year price, q_0 = Base year weighting, q_n = Current year weighting, q_t = Typical year weighting

Laspeyre's Index

$$\frac{\Sigma p_n q_0}{\Sigma p_0 q_0}$$

Paasche's Index

$$\frac{\Sigma p_n q_n}{\Sigma p_0 q_n}$$

Typical Year Index

$$\frac{\Sigma p_n q_t}{\Sigma p_0 q_t}$$

Fisher's Ideal Index

$$\sqrt{\left\{\frac{\Sigma p_n q_0}{\Sigma p_0 q_0}\right\}\left\{\frac{\Sigma p_n q_n}{\Sigma p_0 q_n}\right\}}$$

Marshall Edgeworth Bowley Index

$$\frac{\Sigma p_n(q_0 + q_n)}{\Sigma p_0(q_0 + q_n)}$$

APPENDIX C: Tables

Table 1. Normal distribution tables

a. Table of $\varphi(x)$.

x	$\varphi(x)$	x	$\varphi(x)$	x	$\varphi(x)$	x	$\varphi(x)$
0.0	.399	0.8	.290	1.6	.111	2.4	.022
0.1	.397	0.9	.266	1.7	.094	2.5	.018
0.2	.391	1.0	.242	1.8	.079	2.6	.014
0.3	.381	1.1	.218	1.9	.066	2.7	.010
0.4	.368	1.2	.194	2.0	.054	2.8	.008
0.5	.352	1.3	.171	2.1	.044	2.9	.006
0.6	.333	1.4	.150	2.2	.035	3.0	.004
0.7	.312	1.5	.130	2.3	.028		

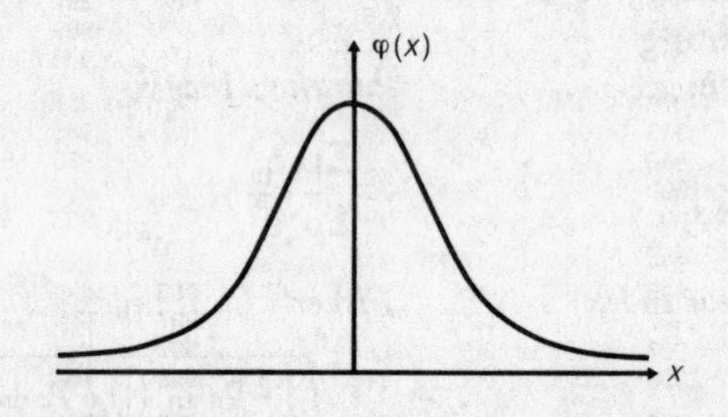

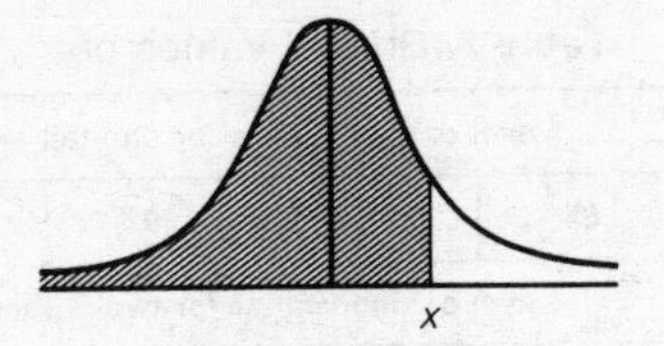

b. Table of $\Phi(x)$.

x	0	1	2	3	4	5	6	7	8	9
0.0	.500	.504	.508	.512	.516	.520	.524	.528	.532	.536
0.1	.540	.544	.548	.552	.556	.560	.564	.567	.571	.575
0.2	.579	.583	.587	.591	.595	.599	.603	.606	.610	.614
0.3	.618	.622	.626	.629	.633	.637	.641	.644	.648	.652
0.4	.655	.659	.663	.666	.670	.674	.677	.681	.684	.688
0.5	.691	.695	.698	.702	.705	.709	.712	.716	.719	.722
0.6	.726	.729	.732	.736	.739	.742	.745	.749	.752	.755
0.7	.758	.761	.764	.767	.770	.773	.776	.779	.782	.785
0.8	.788	.791	.794	.797	.800	.802	.805	.808	.811	.813
0.9	.816	.819	.821	.824	.826	.829	.831	.834	.836	.839
1.0	.841	.844	.846	.848	.851	.853	.855	.858	.860	.862
1.1	.864	.867	.869	.871	.873	.875	.877	.879	.881	.883
1.2	.885	.887	.889	.891	.893	.894	.896	.898	.900	.901
1.3	.903	.905	.907	.908	.910	.911	.913	.915	.916	.918
1.4	.919	.921	.922	.924	.925	.926	.928	.929	.931	.932
1.5	.933	.934	.936	.937	.938	.939	.941	.942	.943	.944
1.6	.945	.946	.947	.948	.949	.951	.952	.953	.954	.954
1.7	.955	.956	.957	.958	.959	.960	.961	.962	.962	.963
1.8	.964	.965	.966	.966	.967	.968	.969	.969	.970	.971
1.9	.971	.972	.973	.973	.974	.974	.975	.976	.976	.977
2.0	.977	.978	.978	.979	.979	.980	.980	.981	.981	.982
2.1	.982	.983	.983	.983	.984	.984	.985	.985	.985	.986
2.2	.986	.986	.987	.987	.987	.988	.988	.988	.989	.989
2.3	.989	.990	.990	.990	.990	.991	.991	.991	.991	.992
2.4	.992	.992	.992	.992	.993	.993	.993	.993	.993	.994
2.5	.994	.994	.994	.994	.994	.995	.995	.995	.995	.995
2.6	.995	.995	.996	.996	.996	.996	.996	.996	.996	.996
2.7	.996	.997	.997	.997	.997	.997	.997	.997	.997	.997
2.8	.997	.997	.997	.997	.997	.997	.997	.998	.998	.998
2.9	.998	.998	.998	.998	.998	.998	.999	.999	.999	.999
3.0	.999	.999	.999	.999	.999	.999	.999	.999	.999	.999

Source: *SMP Advanced Tables*, 3rd edn. 1979 Cambridge University Press.

APPENDIX C

Table 2. Critical values of t.

	Level of significance for one-tail test					
	.10	.05	.025	.01	.005	.0005
	Level of significance for two-tail test					
ν	.20	.10	.05	.02	.01	.001
1	3.078	6.314	12.706	31.821	63.657	636.619
2	1.886	2.920	4.303	6.965	9.925	31.598
3	1.638	2.353	3.182	4.541	5.841	12.941
4	1.533	2.132	2.776	3.747	4.604	8.610
5	1.476	2.015	2.571	3.365	4.032	6.859
6	1.440	1.943	2.447	3.143	3.707	5.959
7	1.415	1.895	2.365	2.998	3.499	5.405
8	1.397	1.860	2.306	2.896	3.355	5.041
9	1.383	1.833	2.262	2.821	3.250	4.781
10	1.372	1.812	2.228	2.764	3.169	4.587
11	1.363	1.796	2.201	2.718	3.106	4.437
12	1.356	1.782	2.179	2.681	3.055	4.318
13	1.350	1.771	2.160	2.650	3.012	4.221
14	1.345	1.761	2.145	2.624	2.977	4.140
15	1.341	1.753	2.131	2.602	2.947	4.073
16	1.337	1.746	2.120	2.583	2.921	4.015
17	1.333	1.740	2.110	2.567	2.898	3.965
18	1.330	1.734	2.101	2.552	2.878	3.922
19	1.328	1.729	2.093	2.539	2.861	3.883
20	1.325	1.725	2.086	2.528	2.845	3.850
21	1.323	1.721	2.080	2.518	2.831	3.819
22	1.321	1.717	2.074	2.508	2.819	3.792
23	1.319	1.714	2.069	2.500	2.807	3.767
24	1.318	1.711	2.064	2.492	2.797	3.745
25	1.316	1.708	2.060	2.485	2.787	3.725
26	1.315	1.706	2.056	2.479	2.779	3.707
27	1.314	1.703	2.052	2.473	2.771	3.690
28	1.313	1.701	2.048	2.467	2.763	3.674
29	1.311	1.699	2.045	2.462	2.756	3.659
30	1.310	1.697	2.042	2.457	2.750	3.646
40	1.303	1.684	2.021	2.423	2.704	3.551
60	1.296	1.671	2.000	2.390	2.460	3.460
120	1.289	1.658	1.980	2.358	2.617	3.373
–	1.282	1.645	1.960	2.326	2.576	3.291

ν = degrees of freedom

Source: *Statistical Tables*, 6th edn. R.A. Fisher & F. Yeates 1974 Longman.

Table 3. Percentage points of the χ^2-distribution

ν \ P	99.5	99	97.5	95	10	5	2.5	1	0.5	0.1
1	0.0^4393	0.0^3157	0.0^3982	0.00393	2.71	3.84	5.02	6.63	7.88	10.83
2	0.0100	0.0201	0.0506	0.103	4.61	5.99	7.38	9.21	10.60	13.81
3	0.0717	0.115	0.216	0.352	6.25	7.81	9.35	11.34	12.84	16.27
4	0.207	0.297	0.484	0.711	7.78	9.49	11.14	13.28	14.86	18.47
5	0.412	0.554	0.831	1.15	9.24	11.07	12.83	15.09	16.75	20.52
6	0.676	0.872	1.24	1.64	10.64	12.59	14.45	16.81	18.55	22.46
7	0.989	1.24	1.69	2.17	12.02	14.07	16.01	18.48	20.28	24.32
8	1.34	1.65	2.18	2.73	13.36	15.51	17.53	20.09	21.95	26.12
9	1.73	2.09	2.70	3.33	14.68	16.92	19.02	21.67	23.59	27.88
10	2.16	2.56	3.25	3.94	15.99	18.31	20.48	23.21	25.19	29.59
11	2.60	3.05	3.82	4.57	17.28	19.68	21.92	24.73	26.76	31.26
12	3.07	3.57	4.40	5.23	18.55	21.03	23.34	26.22	28.30	32.91
13	3.57	4.11	5.01	5.89	19.81	22.36	24.74	27.69	29.82	34.53
14	4.07	4.66	5.63	6.57	21.06	23.68	26.12	29.14	31.32	36.12
15	4.60	5.23	6.26	7.26	22.31	25.00	27.49	30.58	32.80	37.70
16	5.14	5.81	6.91	7.96	23.54	26.30	28.85	32.00	34.27	39.25
17	5.70	6.41	7.56	8.67	24.77	27.59	30.19	33.41	35.72	40.79
18	6.26	7.01	8.23	9.39	25.99	28.87	31.53	34.81	37.16	42.31
19	6.84	7.63	8.91	10.12	27.20	30.14	32.85	36.19	38.58	43.82
20	7.43	8.26	9.59	10.85	28.41	31.41	34.17	37.57	40.00	45.31
21	8.03	8.90	10.28	11.59	29.62	32.67	35.48	38.93	41.40	46.80
22	8.64	9.54	10.98	12.34	30.81	33.92	36.78	40.29	42.80	48.27
23	9.26	10.20	11.69	13.09	32.01	35.17	38.08	41.64	44.18	49.73
24	9.89	10.86	12.40	13.85	33.20	36.42	39.36	42.98	45.56	51.18
25	10.52	11.52	13.12	14.61	34.38	37.65	40.65	44.31	46.93	52.62
26	11.16	12.20	13.84	15.38	35.56	38.89	41.92	45.64	48.29	54.05
27	11.81	12.88	14.57	16.15	36.74	40.11	43.19	46.96	49.64	55.48
28	12.46	13.56	15.31	16.93	37.92	41.34	44.46	48.28	50.99	56.89
29	13.12	14.26	16.05	17.71	39.09	42.56	45.72	49.59	52.34	58.30
30	13.79	14.95	16.79	18.49	40.26	43.77	46.98	50.89	53.67	59.70
40	20.71	22.16	24.43	26.51	51.81	55.76	59.34	63.69	66.77	73.40
50	27.99	29.71	32.36	34.76	63.17	67.50	71.42	76.15	79.49	86.66
60	35.53	37.48	40.48	43.19	74.40	79.08	83.30	88.38	91.95	99.61
70	43.28	45.44	48.76	51.74	85.53	90.53	95.02	100.4	104.2	112.3
80	51.17	53.54	57.15	60.39	96.58	101.9	106.6	112.3	116.3	124.8
90	59.20	61.75	65.65	69.13	107.6	113.1	118.1	124.1	128.3	137.2
100	67.33	70.06	74.22	77.93	118.5	124.3	129.6	135.8	140.2	149.4

Source: *Cambridge Elementary Statistics Tables*, D.V. Lindley & J.C.P. Miller 1953 Cambridge University Press.

Table 4a. 5% points of the *F*-distribution

ν_2 \ ν_1	1	2	3	4	5	6	7	8	10	12	24	∞
1	161.4	199.5	215.7	224.6	230.2	234.0	236.8	238.9	241.9	243.9	249.0	254.3
2	18.5	19.0	19.2	19.2	19.3	19.3	19.4	19.4	19.4	19.4	19.5	19.5
3	10.13	9.55	9.28	9.12	9.01	8.94	8.89	8.85	8.79	8.74	8.64	8.53
4	7.71	6.94	6.59	6.39	6.26	6.16	6.09	6.04	5.96	5.91	5.77	5.63
5	6.61	5.79	5.41	5.19	5.05	4.95	4.88	4.82	4.74	4.68	4.53	4.36
6	5.99	5.14	4.76	4.53	4.39	4.28	4.21	4.15	4.06	4.00	3.84	3.67
7	5.59	4.74	4.35	4.12	3.97	3.87	3.79	3.73	3.64	3.57	3.41	3.23
8	5.32	4.46	4.07	3.84	3.69	3.58	3.50	3.44	3.35	3.28	3.12	2.93
9	5.12	4.26	3.86	3.63	3.48	3.37	3.29	3.23	3.14	3.07	2.90	2.71
10	4.96	4.10	3.71	3.48	3.33	3.22	3.14	3.07	2.98	2.91	2.74	2.54
11	4.84	3.98	3.59	3.36	3.20	3.09	3.01	2.95	2.85	2.79	2.61	2.40
12	4.75	3.89	3.49	3.26	3.11	3.00	2.91	2.85	2.75	2.69	2.51	2.30
13	4.67	3.81	3.41	3.18	3.03	2.92	2.83	2.77	2.67	2.60	2.42	2.21
14	4.60	3.74	3.34	3.11	2.96	2.85	2.76	2.70	2.60	2.53	2.35	2.13
15	4.54	3.68	3.29	3.06	2.90	2.79	2.71	2.64	2.54	2.48	2.29	2.07
16	4.49	3.63	3.24	3.01	2.85	2.74	2.66	2.59	2.49	2.42	2.24	2.01
17	4.45	3.59	3.20	2.96	2.81	2.70	2.61	2.55	2.45	2.38	2.19	1.96
18	4.41	3.55	3.16	2.93	2.77	2.66	2.58	2.51	2.41	2.34	2.15	1.92
19	4.38	3.52	3.13	2.90	2.74	2.63	2.54	2.48	2.38	2.31	2.11	1.88
20	4.35	3.49	3.10	2.87	2.71	2.60	2.51	2.45	2.35	2.28	2.08	1.84
21	4.32	3.47	3.07	2.84	2.68	2.57	2.49	2.42	2.32	2.25	2.05	1.81
22	4.30	3.44	3.05	2.82	2.66	2.55	2.46	2.40	2.30	2.23	2.03	1.78
23	4.28	3.42	3.03	2.80	2.64	2.53	2.44	2.37	2.27	2.20	2.00	1.76
24	4.26	3.40	3.01	2.78	2.62	2.51	2.42	2.36	2.25	2.18	1.98	1.73
25	4.24	3.39	2.99	2.76	2.60	2.49	2.40	2.34	2.24	2.16	1.96	1.71
26	4.23	3.37	2.98	2.74	2.59	2.47	2.39	2.32	2.22	2.15	1.95	1.69
27	4.21	3.35	2.96	2.73	2.57	2.46	2.37	2.31	2.20	2.13	1.93	1.67
28	4.20	3.34	2.95	2.71	2.56	2.45	2.36	2.29	2.19	2.12	1.91	1.65
29	4.18	3.33	2.93	2.70	2.55	2.43	2.35	2.28	2.18	2.10	1.90	1.64
30	4.17	3.32	2.92	2.69	2.53	2.42	2.33	2.27	2.16	2.09	1.89	1.62
32	4.15	3.29	2.90	2.67	2.51	2.40	2.31	2.24	2.14	2.07	1.86	1.59
34	4.13	3.28	2.88	2.65	2.49	2.38	2.29	2.23	2.12	2.05	1.84	1.57
36	4.11	3.26	2.87	2.63	2.48	2.36	2.28	2.21	2.11	2.03	1.82	1.55
38	4.10	3.24	2.85	2.62	2.46	2.35	2.26	2.19	2.09	2.02	1.81	1.53
40	4.08	3.23	2.84	2.61	2.45	2.34	2.25	2.18	2.08	2.00	1.79	1.51
60	4.00	3.15	2.76	2.53	2.37	2.25	2.17	2.10	1.99	1.92	1.70	1.39
120	3.92	3.07	2.68	2.45	2.29	2.18	2.09	2.02	1.91	1.83	1.61	1.25
∞	3.84	3.00	2.60	2.37	2.21	2.10	2.01	1.94	1.83	1.75	1.52	1.00

Source: *Cambridge Elementary Statistics Tables,* D.V. Lindley & J.C.P. Miller 1953 Cambridge University Press

Table 4b. $2\frac{1}{2}$% points of the *F*-distribution

ν_2 \ ν_1	1	2	3	4	5	6	7	8	10	12	24	∞
1	648	800	864	900	922	937	948	957	969	977	997	1018
2	38.5	39.0	39.2	39.2	39.3	39.3	39.4	39.4	39.4	39.4	39.5	39.5
3	17.4	16.0	15.4	15.1	14.9	14.7	14.6	14.5	14.4	14.3	14.1	13.9
4	12.22	10.65	9.98	9.60	9.36	9.20	9.07	8.98	8.84	8.75	8.51	8.26
5	10.01	8.43	7.76	7.39	7.15	6.98	6.85	6.76	6.62	6.52	6.28	6.02
6	8.81	7.26	6.60	6.23	5.99	5.82	5.70	5.60	5.46	5.37	5.12	4.85
7	8.07	6.54	5.89	5.52	5.29	5.12	4.99	4.90	4.76	4.67	4.42	4.14
8	7.57	6.06	5.42	5.05	4.82	4.65	4.53	4.43	4.30	4.20	3.95	3.67
9	7.21	5.71	5.08	4.72	4.48	4.32	4.20	4.10	3.96	3.87	3.61	3.33
10	6.94	5.46	4.83	4.47	4.24	4.07	3.95	3.85	3.72	3.62	3.37	3.08
11	6.72	5.26	4.63	4.28	4.04	3.88	3.76	3.66	3.53	3.43	3.17	2.88
12	6.55	5.10	4.47	4.12	3.89	3.73	3.61	3.51	3.37	3.28	3.02	2.72
13	6.41	4.97	4.35	4.00	3.77	3.60	3.48	3.39	3.25	3.15	2.89	2.60
14	6.30	4.86	4.24	3.89	3.66	3.50	3.38	3.29	3.15	3.05	2.79	2.49
15	6.20	4.76	4.15	3.80	3.58	3.41	3.29	3.20	3.06	2.96	2.70	2.40
16	6.12	4.69	4.08	3.73	3.50	3.34	3.22	3.12	2.99	2.89	2.63	2.32
17	6.04	4.62	4.01	3.66	3.44	3.28	3.16	3.06	2.92	2.82	2.56	2.25
18	5.98	4.56	3.95	3.61	3.38	3.22	3.10	3.01	2.87	2.77	2.50	2.19
19	5.92	4.51	3.90	3.56	3.33	3.17	3.05	2.96	2.82	2.72	2.45	2.13
20	5.87	4.46	3.86	3.51	3.29	3.13	3.01	2.91	2.77	2.68	2.41	2.09
21	5.83	4.42	3.82	3.48	3.25	3.09	2.97	2.87	2.73	2.64	2.37	2.04
22	5.79	4.38	3.78	3.44	3.22	3.05	2.93	2.84	2.70	2.60	2.33	2.00
23	5.75	4.35	3.75	3.41	3.18	3.02	2.90	2.81	2.67	2.57	2.30	1.97
24	5.72	4.32	3.72	3.38	3.15	2.99	2.87	2.78	2.64	2.54	2.27	1.94
25	5.69	4.29	3.69	3.35	3.13	2.97	2.85	2.75	2.61	2.51	2.24	1.91
26	5.66	4.27	3.67	3.33	3.10	2.94	2.82	2.73	2.59	2.49	2.22	1.88
27	5.63	4.24	3.65	3.31	3.08	2.92	2.80	2.71	2.57	2.47	2.19	1.85
28	5.61	4.22	3.63	3.29	3.06	2.90	2.78	2.69	2.55	2.45	2.17	1.83
29	5.59	4.20	3.61	3.27	3.04	2.88	2.76	2.67	2.53	2.43	2.15	1.81
30	5.57	4.18	3.59	3.25	3.03	2.87	2.75	2.65	2.51	2.41	2.14	1.79
32	5.53	4.15	3.56	3.22	3.00	2.84	2.72	2.62	2.48	2.38	2.10	1.75
34	5.50	4.12	3.53	3.19	2.97	2.81	2.69	2.59	2.45	2.35	2.08	1.72
36	5.47	4.09	3.51	3.17	2.94	2.79	2.66	2.57	2.43	2.33	2.05	1.69
38	5.45	4.07	3.48	3.15	2.92	2.76	2.64	2.55	2.41	2.31	2.03	1.66
40	5.42	4.05	3.46	3.13	2.90	2.74	2.62	2.53	2.39	2.29	2.01	1.64
60	5.29	3.93	3.34	3.01	2.79	2.63	2.51	2.41	2.27	2.17	1.88	1.48
120	5.15	3.80	3.23	2.89	2.67	2.52	2.39	2.30	2.16	2.05	1.76	1.31
∞	5.02	3.69	3.12	2.79	2.57	2.41	2.29	2.19	2.05	1.94	1.64	1.00

Table 4c. 1% points of the *F*-distribution

ν_2 \ ν_1	1	2	3	4	5	6	7	8	10	12	24	∞
1	4052	5000	5403	5625	5764	5859	5928	5981	6056	6106	6235	6366
2	98.5	99.0	99.2	99.2	99.3	99.3	99.4	99.4	99.4	99.4	99.5	99.5
3	34.1	30.8	29.5	28.7	28.2	27.9	27.7	27.5	27.2	27.1	26.6	26.1
4	21.2	18.0	16.7	16.0	15.5	15.2	15.0	14.8	14.5	14.4	13.9	13.5
5	16.26	13.27	12.06	11.39	10.97	10.67	10.46	10.29	10.05	9.89	9.47	9.02
6	13.74	10.92	9.78	9.15	8.75	8.47	8.26	8.10	7.87	7.72	7.31	6.88
7	12.25	9.55	8.45	7.85	7.46	7.19	6.99	6.84	6.62	6.47	6.07	5.65
8	11.26	8.65	7.59	7.01	6.63	6.37	6.18	6.03	5.81	5.67	5.28	4.86
9	10.56	8.02	6.99	6.42	6.06	5.80	5.61	5.47	5.26	5.11	4.73	4.31
10	10.04	7.56	6.55	5.99	5.64	5.39	5.20	5.06	4.85	4.71	4.33	3.91
11	9.65	7.21	6.22	5.67	5.32	5.07	4.89	4.74	4.54	4.40	4.02	3.60
12	9.33	6.93	5.95	5.41	5.06	4.82	4.64	4.50	4.30	4.16	3.78	3.36
13	9.07	6.70	5.74	5.21	4.86	4.62	4.44	4.30	4.10	3.96	3.59	3.17
14	8.86	6.51	5.56	5.04	4.70	4.46	4.28	4.14	3.94	3.80	3.43	3.00
15	8.68	6.36	5.42	4.89	4.56	4.32	4.14	4.00	3.80	3.67	3.29	2.87
16	8.53	6.23	5.29	4.77	4.44	4.20	4.03	3.89	3.69	3.55	3.18	2.75
17	8.40	6.11	5.18	4.67	4.34	4.10	3.93	3.79	3.59	3.46	3.08	2.65
18	8.29	6.01	5.09	4.58	4.25	4.01	3.84	3.71	3.51	3.37	3.00	2.57
19	8.18	5.93	5.01	4.50	4.17	3.94	3.77	3.63	3.43	3.30	2.92	2.49
20	8.10	5.85	4.94	4.43	4.10	3.87	3.70	3.56	3.37	3.23	2.86	2.42
21	8.02	5.78	4.87	4.37	4.04	3.81	3.64	3.51	3.31	3.17	2.80	2.36
22	7.95	5.72	4.82	4.31	3.99	3.76	3.59	3.45	3.26	3.12	2.75	2.31
23	7.88	5.66	4.76	4.26	3.94	3.71	3.54	3.41	3.21	3.07	2.70	2.26
24	7.82	5.61	4.72	4.22	3.90	3.67	3.50	3.36	3.17	3.03	2.66	2.21
25	7.77	5.57	4.68	4.18	3.86	3.63	3.46	3.32	3.13	2.99	2.62	2.17
26	7.72	5.53	4.64	4.14	3.82	3.59	3.42	3.29	3.09	2.96	2.58	2.13
27	7.68	5.49	4.60	4.11	3.78	3.56	3.39	3.26	3.06	2.93	2.55	2.10
28	7.64	5.45	4.57	4.07	3.75	3.53	3.36	3.23	3.03	2.90	2.52	2.06
29	7.60	5.42	4.54	4.04	3.73	3.50	3.33	3.20	3.00	2.87	2.49	2.03
30	7.56	5.39	4.51	4.02	3.70	3.47	3.30	3.17	2.98	2.84	2.47	2.01
32	7.50	5.34	4.46	3.97	3.65	3.43	3.26	3.13	2.93	2.80	2.42	1.96
34	7.45	5.29	4.42	3.93	3.61	3.39	3.22	3.09	2.90	2.76	2.38	1.91
36	7.40	5.25	4.38	3.89	3.58	3.35	3.18	3.05	2.86	2.72	2.35	1.87
38	7.35	5.21	4.34	3.86	3.54	3.32	3.15	3.02	2.83	2.69	2.32	1.84
40	7.31	5.18	4.31	3.83	3.51	3.29	3.12	2.99	2.80	2.66	2.29	1.80
60	7.08	4.98	4.13	3.65	3.34	3.12	2.95	2.82	2.63	2.50	2.12	1.60
120	6.85	4.79	3.95	3.48	3.17	2.96	2.79	2.66	2.47	2.34	1.95	1.38
∞	6.63	4.61	3.78	3.32	3.02	2.80	2.64	2.51	2.32	2.18	1.79	1.00

Table 4d. 0.1% points of the *F*-distribution

ν_2 \ ν_1	1	2	3	4	5	6	7	8	10	12	24	∞
1	4053	5000	5404	5625	5764	5859	5929	5981	6056	6107	6235	6366
2	998.5	999.0	999.2	999.2	999.3	999.3	999.4	999.4	999.4	999.4	999.5	999.5
3	167.0	148.5	141.1	137.1	134.6	132.8	131.5	130.6	129.2	128.3	125.9	123.5
4	74.14	61.25	56.18	53.44	51.71	50.53	49.66	49.00	48.05	47.41	45.77	44.05
5	47.18	37.12	33.20	31.09	29.75	28.83	28.16	27.65	26.92	26.42	25.14	23.79
6	35.51	27.00	23.70	21.92	20.80	20.03	19.46	19.03	18.41	17.99	16.90	15.75
7	29.25	21.69	18.77	17.20	16.21	15.52	15.02	14.63	14.08	13.71	12.73	11.70
8	25.42	18.49	15.83	14.39	13.48	12.86	12.40	12.05	11.54	11.19	10.30	9.34
9	22.86	16.39	13.90	12.56	11.71	11.13	10.69	10.37	9.87	9.57	8.72	7.81
10	21.04	14.91	12.55	11.28	10.48	9.93	9.52	9.20	8.74	8.44	7.64	6.76
11	19.69	13.81	11.56	10.35	9.58	9.05	8.66	8.35	7.92	7.63	6.85	6.00
12	18.64	12.97	10.80	9.63	8.89	8.38	8.00	7.71	7.29	7.00	6.25	5.42
13	17.82	12.31	10.21	9.07	8.35	7.86	7.49	7.21	6.80	6.52	5.78	4.97
14	17.14	11.78	9.73	8.62	7.92	7.44	7.08	6.80	6.40	6.13	5.41	4.60
15	16.59	11.34	9.34	8.25	7.57	7.09	6.74	6.47	6.08	5.81	5.10	4.31
16	16.12	10.97	9.01	7.94	7.27	6.80	6.46	6.19	5.81	5.55	4.85	4.06
17	15.72	10.66	8.73	7.68	7.02	6.56	6.22	5.96	5.58	5.32	4.63	3.85
18	15.38	10.39	8.49	7.46	6.81	6.35	6.02	5.76	5.39	5.13	4.45	3.67
19	15.08	10.16	8.28	7.27	6.62	6.18	5.85	5.59	5.22	4.97	4.29	3.51
20	14.82	9.95	8.10	7.10	6.46	6.02	5.69	5.44	5.08	4.82	4.15	3.38
21	14.59	9.77	7.94	6.95	6.32	5.88	5.56	5.31	4.95	4.70	4.03	3.26
22	14.38	9.61	7.80	6.81	6.19	5.76	5.44	5.19	4.83	4.58	3.92	3.15
23	14.19	9.47	7.67	6.70	6.08	5.65	5.33	5.09	4.73	4.48	3.82	3.05
24	14.03	9.34	7.55	6.59	5.98	5.55	5.23	4.99	4.64	4.39	3.74	2.97
25	13.88	9.22	7.45	6.49	5.89	5.46	5.15	4.91	4.56	4.31	3.66	2.89
26	13.74	9.12	7.36	6.41	5.80	5.38	5.07	4.83	4.48	4.24	3.59	2.82
27	13.61	9.02	7.27	6.33	5.73	5.31	5.00	4.76	4.41	4.17	3.52	2.75
28	13.50	8.93	7.19	6.25	5.66	5.24	4.93	4.69	4.35	4.11	3.46	2.69
29	13.39	8.85	7.12	6.19	5.59	5.18	4.87	4.64	4.29	4.05	3.41	2.64
30	13.29	8.77	7.05	6.12	5.53	5.12	4.82	4.58	4.24	4.00	3.36	2.59
32	13.12	8.64	6.94	6.01	5.43	5.02	4.72	4.48	4.14	3.91	3.27	2.50
34	12.97	8.52	6.83	5.92	5.34	4.93	4.63	4.40	4.06	3.83	3.19	2.42
36	12.83	8.42	6.74	5.84	5.26	4.86	4.56	4.33	3.99	3.76	3.12	2.35
38	12.71	8.33	6.66	5.76	5.19	4.79	4.49	4.26	3.93	3.70	3.06	2.29
40	12.61	8.25	6.59	5.70	5.13	4.73	4.44	4.21	3.87	3.64	3.01	2.23
60	11.97	7.77	6.17	5.31	4.76	4.37	4.09	3.86	3.54	3.32	2.69	1.89
120	11.38	7.32	5.78	4.95	4.42	4.04	3.77	3.55	3.24	3.02	2.40	1.54
∞	10.83	6.91	5.42	4.62	4.10	3.74	3.47	3.27	2.96	2.74	2.13	1.00

Table 5. Critical values of the Pearson product moment correlation coefficient.

ν	Level of significance for two-tail test			
	.10	.05	.02	.01
1	.988	.997	$.9^{3}51$	$.9^{3}88$
2	.900	.950	.980	.990
3	.805	.878	.934	.959
4	.729	.811	.882	.917
5	.669	.755	.833	.875
6	.622	.707	.789	.834
7	.582	.666	.750	.798
8	.549	.632	.716	.765
9	.521	.602	.685	.735
10	.497	.576	.658	.708
12	.458	.532	.612	.661
14	.426	.497	.574	.623
16	.400	.468	.543	.590
18	.378	.444	.516	.561
20	.360	.423	.492	.537
25	.323	.381	.445	.487
30	.296	.349	.409	.449
40	.257	.304	.358	.393
50	.231	.273	.322	.354
100	.164	.195	.230	.254

Source: *Statistical Tables for Students*, J.S. Fowlie, Oliver & Boyd.

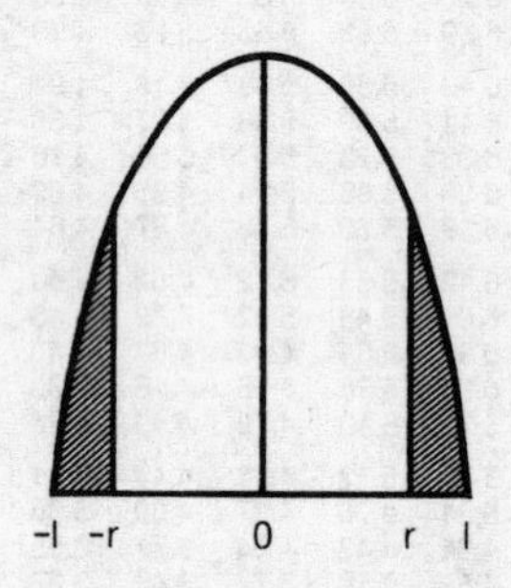

Table 6. Critical values for Spearman's rank correlation coefficient.

	Level of significance for one tail test			
	.05	.025	.01	.005
	Level of significance for two-tail test			
ν	.10	.05	.02	.01
3	.900	1.000	1.000	–
4	.829	.886	.943	1.000
5	.714	.786	.893	.929
6	.643	.738	.833	.881
7	.600	.683	.783	.833
8	.564	.648	.746	.794
10	.506	.591	.712	.777
12	.456	.544	.645	.715
14	.425	.506	.601	.665
16	.399	.475	.564	.625
18	.377	.450	.534	.591
20	.359	.428	.508	.562
22	.343	.409	.485	.537
24	.329	.392	.465	.515
26	.317	.377	.448	.496
28	.306	.364	.432	.478

Table 7. Critical values for Kendall's coefficient of rank correlation.

	Level of significance for one-tail test			
	.05	.025	.01	.005
	Level of significance for two-tail test			
ν	.10	.05	.02	.01
1	–	–	–	–
2	1.0000	–	–	–
3	0.8000	1.0000	1.0000	–
4	0.7333	0.8667	0.8667	1.0000
5	0.6190	0.7143	0.8095	0.9048
6	0.5714	0.6429	0.7143	0.7857
7	0.5000	0.5556	0.6667	0.7222
8	0.4667	0.5111	0.6000	0.6444
9	0.4182	0.4909	0.5636	0.6000
10	0.3939	0.4545	0.5455	0.5758
11	0.3590	0.4359	0.5128	0.5641
12	0.3626	0.4066	0.4725	0.5165
13	0.3333	0.3905	0.4667	0.5048
14	0.3167	0.3833	0.4333	0.4833
15	0.3088	0.3676	0.4265	0.4706
16	0.2941	0.3464	0.4118	0.4510
17	0.2666	0.3333	0.3918	0.4386
18	0.2737	0.3263	0.3789	0.4211
19	0.2867	0.3143	0.3714	0.4095
20	0.2641	0.3074	0.3593	0.3939
21	0.2569	0.2964	0.3518	0.3913
22	0.2464	0.2899	0.3406	0.3768
23	0.2400	0.2867	0.3333	0.3667
24	0.2369	0.2800	0.3292	0.3600
25	0.2308	0.2707	0.3219	0.3561
26	0.2275	0.2646	0.3122	0.3439
27	0.2217	0.2611	0.3103	0.3399
28	0.2184	0.2552	0.3011	0.3333
29	0.2129	0.2516	0.2946	0.3247
30	0.2097	0.2460	0.2903	0.3226

Source: *Elementary Statistics Tables*, H.R. Neave 1981, George Allen & Unwin.

Table 8. Fisher's z-transformation of r to z_r.

r	z_r	r	z_r	r	z_r
.01	.010	.34	.354	.67	.811
.02	.020	.35	.366	.68	.829
.03	.030	.36	.377	.69	.848
.04	.040	.37	.389	.70	.867
.05	.050	.38	.400	.71	.887
.06	.060	.39	.412	.72	.908
.07	.070	.40	.424	.73	.929
.08	.080	.41	.436	.74	.950
.09	.090	.42	.448	.75	.973
.10	.100	.43	.460	.76	.996
.11	.110	.44	.472	.77	1.020
.12	.121	.45	.485	.78	1.045
.13	.131	.46	.497	.79	1.071
.14	.141	.47	.510	.80	1.099
.15	.151	.48	.523	.81	1.127
.16	.161	.49	.536	.82	1.157
.17	.172	.50	.549	.83	1.188
.18	.181	.51	.563	.84	1.221
.19	.192	.52	.577	.85	1.256
.20	.203	.53	.590	.86	1.293
.21	.214	.54	.604	.87	1.333
.22	.224	.55	.618	.88	1.376
.23	.234	.56	.633	.89	1.422
.24	.245	.57	.648	.90	1.472
.25	.256	.58	.663	.91	1.528
.26	.266	.59	.678	.92	1.589
.27	.277	.60	.693	.93	1.658
.28	.288	.61	.709	.94	1.738
.29	.299	.62	.725	.95	1.832
.30	.309	.63	.741	.96	1.946
.31	.321	.64	.758	.97	2.092
.32	.332	.65	.775	.98	2.298
.33	.343	.66	.793	.99	2.647

Source: *Psychology Statistics*, Q. McNemar, John Wiley & Sons, Inc.

Table 9. Conversion of range to standard deviation.

n	a_n	n	a_n	n	a_n	n	a_n
2	0.8862	5	0.4299	8	0.3512	11	0.3152
3	0.5908	6	0.3946	9	0.3367	12	0.3069
4	0.4857	7	0.3698	10	0.3249	13	0.2998

Source: *Advanced General Statistics*, B.C. Erricker 1971, Hodder & Stoughton.

APPENDIX C

Table 10a. Critical values of U_1 and U_2 for a one-tail test at 2.5% significance level, or a two-tail test at 5% significance level

n_2 \ n_1	1	2	3	4	5	6	7	8	9	10	11	12	13	14	15	16	17	18	19	20
1	–	–	–	–	–	–	–	–	–	–	–	–	–	–	–	–	–	–	–	–
2	–	–	–	–	–	–	–	0 16	0 18	0 20	0 22	1 23	1 25	1 27	1 29	1 31	2 32	2 34	2 36	2 38
3	–	–	–	–	0 15	1 17	1 20	2 22	2 25	3 27	3 30	4 32	4 35	5 37	5 40	6 42	6 45	7 47	7 50	8 52
4	–	–	–	0 16	1 19	2 22	3 25	4 28	4 32	5 35	6 38	7 41	8 44	9 47	10 50	11 53	11 57	12 60	13 63	13 67
5	–	–	0 15	1 19	2 23	3 27	5 30	6 34	7 38	8 42	9 46	11 49	12 53	13 57	14 61	15 65	17 68	18 72	19 76	20 80
6	–	–	1 17	2 22	3 27	5 31	6 36	8 40	10 44	11 49	13 53	14 58	16 62	17 67	19 71	21 75	22 80	24 84	25 89	27 93
7	–	–	1 20	3 25	5 30	6 36	8 41	10 46	12 51	14 56	16 61	18 66	20 71	22 76	24 81	26 86	28 91	30 96	32 101	34 106
8	–	0 16	2 22	4 28	6 34	8 40	10 46	13 51	15 57	17 63	19 69	22 74	24 80	26 86	29 91	31 97	34 102	36 108	38 111	41 119
9	–	0 18	2 25	4 32	7 38	10 44	12 51	15 57	17 64	20 70	23 76	26 82	28 89	31 95	34 101	37 107	39 114	42 120	45 126	48 132
10	–	0 20	3 27	5 35	8 42	11 49	14 56	17 63	20 70	23 77	26 84	29 91	33 97	36 104	39 111	42 118	45 125	48 132	52 138	55 145
11	–	0 22	3 30	6 38	9 46	13 53	16 61	19 69	23 76	26 84	30 91	33 99	37 106	40 114	44 121	47 129	51 136	55 143	58 151	62 158
12	–	1 23	4 32	7 41	11 49	14 58	18 66	22 74	26 82	29 91	33 99	37 107	41 115	45 123	49 131	53 139	57 147	61 155	65 163	69 171
13	–	1 25	4 35	8 44	12 53	16 62	20 71	24 80	28 89	33 97	37 106	41 115	45 124	50 132	54 141	59 149	63 158	67 167	72 175	76 184
14	–	1 27	5 37	9 47	13 51	17 67	22 76	26 86	31 95	36 104	40 114	45 123	50 132	55 141	59 151	64 160	67 171	74 178	78 188	83 197
15	–	1 29	5 40	10 50	14 61	19 71	24 81	29 91	34 101	39 111	44 121	49 131	54 141	59 151	64 161	70 170	75 180	80 190	85 200	90 210
16	–	1 31	6 42	11 53	15 65	21 75	26 86	31 97	37 107	42 118	47 129	53 139	59 149	64 160	70 170	75 181	81 191	86 202	92 212	98 222
17	–	2 32	6 45	11 57	17 68	22 80	28 91	34 102	39 114	45 125	51 136	57 147	63 158	67 171	75 180	81 191	87 202	93 213	99 224	105 235
18	–	2 34	7 47	12 60	18 72	24 84	30 96	36 108	42 120	48 132	55 143	61 155	67 167	74 178	80 190	86 202	93 213	99 225	106 236	112 248
19	–	2 36	7 50	13 63	19 76	25 89	32 101	38 114	45 126	52 138	58 151	65 163	72 175	78 188	85 200	92 212	99 224	106 236	113 248	119 261
20	–	2 38	8 52	13 67	20 80	27 93	34 106	41 119	48 132	55 145	62 158	69 171	76 184	83 197	90 210	98 222	105 235	112 248	119 261	127 273

(Dashes in the body of the table indicate that no decision is possible at the stated level of significance.) Source: *Statistics – An Interdisciplinary Approach*. Open University.

Table 10b. Critical values of U_1 and U_2 for a one-tail test at 5% significance level, or a two-tail test at 10% significance level.

n_2 \ n_1	1	2	3	4	5	6	7	8	9	10	11	12	13	14	15	16	17	18	19	20
1	–	–	–	–	–	–	–	–	–	–	–	–	–	–	–	–	–	–	0 19	0 20
2	–	–	–	–	0 10	0 12	0 14	1 15	1 17	1 19	1 21	2 22	2 24	2 26	3 27	3 29	3 31	4 32	4 34	4 36
3	–	–	0 9	0 12	1 14	2 16	2 19	3 21	3 24	4 26	5 28	5 31	6 33	7 35	7 38	8 40	9 42	9 45	10 47	11 49
4	–	–	0 12	1 15	2 18	3 21	4 24	5 27	6 30	7 33	8 36	9 39	10 42	11 45	12 48	14 50	15 53	16 56	17 59	18 62
5	–	0 10	1 14	2 18	4 21	5 25	6 29	8 32	9 36	11 39	12 43	13 47	15 50	16 54	18 57	19 61	20 65	22 68	23 72	25 75
6	–	0 12	2 16	3 21	5 25	7 29	8 34	10 38	12 42	14 46	16 50	17 55	19 59	21 63	23 67	25 71	26 76	28 80	30 84	32 88
7	–	0 14	2 19	4 24	6 29	8 34	11 38	13 43	15 48	17 53	19 58	21 63	24 67	26 72	28 77	30 82	33 86	35 91	37 96	39 101
8	–	1 15	3 21	5 27	8 32	10 38	13 43	15 49	18 54	20 60	23 65	26 70	28 76	31 81	33 87	36 92	39 97	41 103	44 108	47 113
9	–	1 17	3 24	6 30	9 36	12 42	15 48	18 54	21 60	24 66	27 72	30 78	33 84	36 90	39 96	42 102	45 108	48 114	51 120	54 126
10	–	1 19	4 26	7 33	11 39	14 46	17 53	20 60	24 66	27 73	31 79	34 86	37 93	41 99	44 106	48 112	51 119	55 125	58 132	62 138
11	–	1 21	5 28	8 36	12 43	16 50	19 58	23 65	27 72	31 79	34 87	38 94	42 101	46 108	50 115	54 122	57 130	61 137	65 144	69 151
12	–	2 22	5 31	9 39	13 47	17 55	21 63	26 70	30 78	34 86	38 94	42 102	47 109	51 117	55 125	60 132	64 140	68 148	72 156	77 163
13	–	2 24	6 33	10 42	15 50	19 59	24 67	28 76	33 84	37 93	42 101	47 109	51 118	56 126	61 134	65 143	70 151	75 159	80 167	84 176
14	–	2 26	7 35	11 45	16 54	21 63	26 72	31 81	36 90	41 99	46 108	51 117	56 126	61 135	66 144	71 153	77 161	82 170	87 179	92 188
15	–	3 27	7 38	12 48	18 57	23 67	28 77	33 87	39 96	44 106	50 115	55 125	61 134	66 144	72 153	77 163	83 172	88 182	94 191	100 200
16	–	3 29	8 40	14 50	19 61	25 71	30 82	36 92	42 102	48 112	54 122	60 132	65 143	71 153	77 163	83 173	89 183	95 193	101 203	107 213
17	–	3 31	9 42	15 53	20 65	26 76	33 86	39 97	45 108	51 119	57 130	64 140	70 151	77 161	83 172	89 183	96 193	102 204	109 214	115 225
18	–	4 32	9 45	16 56	22 68	28 80	35 91	41 103	48 114	55 123	61 137	68 148	75 159	82 170	88 182	95 193	102 204	109 215	116 226	123 237
19	0 19	4 34	10 47	17 59	23 72	30 84	37 96	44 108	51 120	58 132	65 144	72 156	80 167	87 179	94 191	101 203	109 214	116 226	123 238	130 250
20	0 20	4 36	11 49	18 62	25 75	32 88	39 101	47 113	54 126	62 138	69 151	77 163	84 176	92 188	100 200	107 213	115 225	123 237	130 250	138 262

(Dashes in the body of the table indicate that no decision is possible at the stated level of significance.)

Table 10c. Critical values of U_1 and U_2 for a one-tail test at 0.5% significance level, or a two-tail test at 1% significance level.

n_2 \ n_1	1	2	3	4	5	6	7	8	9	10	11	12	13	14	15	16	17	18	19	20
1	–	–	–	–	–	–	–	–	–	–	–	–	–	–	–	–	–	–	–	–
2	–	–	–	–	–	–	–	–	–	–	–	–	–	–	–	–	–	–	0 38	0 40
3	–	–	–	–	–	–	–	–	0 27	0 30	0 33	1 35	1 38	1 41	2 43	2 46	2 49	2 52	3 54	3 57
4	–	–	–	–	–	0 24	0 28	1 31	1 35	2 38	2 42	3 45	3 49	4 52	5 55	5 59	6 62	6 66	7 69	8 72
5	–	–	–	–	0 25	1 29	1 34	2 38	3 42	4 46	5 50	6 54	7 58	7 63	8 67	9 71	10 75	11 79	12 83	13 87
6	–	–	–	0 24	1 29	2 34	3 39	4 44	5 49	6 54	7 59	9 63	10 68	11 73	12 78	13 83	15 87	16 92	17 97	18 102
7	–	–	–	0 28	1 34	3 39	4 45	6 50	7 56	9 61	10 67	12 72	13 78	15 83	16 89	18 94	19 100	21 105	22 111	24 116
8	–	–	–	1 31	2 38	4 44	6 50	7 57	9 63	11 69	13 75	15 81	17 87	18 94	20 100	22 106	24 112	26 118	28 124	30 130
9	–	–	0 27	1 35	3 42	5 49	7 56	9 63	11 70	13 77	16 83	18 90	20 97	22 104	24 111	27 117	29 124	31 131	33 138	36 144
10	–	–	0 30	2 38	4 46	6 54	9 61	11 69	13 77	16 84	18 92	21 99	24 106	26 114	29 121	31 129	34 136	37 143	39 151	42 158
11	–	–	0 33	2 42	5 50	7 59	10 67	13 75	16 83	18 92	21 100	24 108	27 116	30 124	33 132	36 140	39 148	42 156	45 164	48 172
12	–	–	1 36	3 45	6 54	9 63	12 72	15 81	18 90	21 99	24 108	27 117	31 125	34 134	37 143	41 151	44 160	47 169	51 177	54 186
13	–	–	1 38	3 49	7 58	10 68	13 78	17 87	20 97	24 106	27 116	31 125	34 125	38 144	42 153	45 163	49 172	53 181	56 191	60 200
14	–	–	1 41	4 52	7 63	11 73	15 83	18 94	22 104	26 114	30 124	34 134	38 144	42 154	46 164	50 174	54 184	58 194	63 203	67 213
15	–	–	2 43	5 55	8 67	12 78	16 89	20 100	24 111	29 121	33 132	37 143	42 153	46 161	51 174	55 185	60 195	64 206	69 216	73 227
16	–	–	2 46	5 59	9 71	13 83	18 94	22 106	27 117	31 129	36 140	41 151	45 163	50 174	55 185	60 196	65 207	70 218	74 230	79 241
17	–	–	2 49	6 62	10 75	15 87	19 100	24 112	29 124	34 148	39 148	44 160	49 172	54 184	60 195	65 207	70 219	75 231	81 242	86 254
18	–	–	2 52	6 66	11 79	16 92	21 105	26 118	31 131	37 143	42 156	47 169	53 181	58 194	64 206	70 218	75 231	81 243	87 255	92 268
19	–	0 38	3 54	7 69	12 83	17 97	22 111	28 124	33 138	39 151	45 164	51 177	56 191	63 203	69 216	74 230	81 242	87 255	93 268	99 281
20	–	0 40	3 57	8 72	13 87	18 102	24 116	30 130	36 144	42 158	48 172	54 186	60 200	67 213	73 227	79 241	86 254	92 268	99 281	105 295

(Dashes in the body of the table indicate that no decision is possible at the stated level of significance.)

Table 10d. Critical values of U_1 and U_2 for a one-tail test at 1% significance level, or a two-tail test at 2% significance level.

n_2 \ n_1	1	2	3	4	5	6	7	8	9	10	11	12	13	14	15	16	17	18	19	20
1	–	–	–	–	–	–	–	–	–	–	–	–	–	–	–	–	–	–	–	–
2	–	–	–	–	–	–	–	–	–	–	–	–	0 26	0 28	0 30	0 32	0 34	0 36	1 37	1 39
3	–	–	–	–	–	–	0 21	0 24	1 26	1 29	1 32	2 34	2 37	2 40	3 42	3 45	4 47	4 50	4 52	5 55
4	–	–	–	–	0 20	1 23	1 27	2 30	3 33	3 37	4 40	5 43	5 47	6 50	7 53	7 57	8 60	9 63	9 67	10 70
5	–	–	–	0 20	1 24	2 28	3 32	4 36	5 40	6 44	7 48	8 52	9 56	10 60	11 64	12 68	13 72	14 76	15 80	16 84
6	–	–	–	1 23	2 28	3 33	4 38	6 42	7 47	8 52	9 57	11 61	12 66	13 71	15 75	16 80	18 84	19 89	20 94	22 98
7	–	–	0 21	1 27	3 32	4 38	6 43	7 49	9 54	11 59	12 65	14 70	16 75	17 81	19 86	21 91	23 96	24 102	26 107	28 112
8	–	–	0 24	2 30	4 36	6 42	7 49	9 55	11 61	13 67	15 73	17 79	20 84	22 90	24 96	26 102	28 108	30 114	32 120	34 126
9	–	–	1 26	3 33	5 40	7 47	9 54	11 61	14 67	16 74	18 81	21 87	23 94	26 100	28 107	31 113	33 120	36 126	38 133	40 140
10	–	–	1 29	3 37	6 44	8 52	11 59	13 67	16 74	19 81	22 88	24 96	27 103	30 110	33 117	36 124	38 132	41 139	44 146	47 153
11	–	–	1 32	4 40	7 48	9 57	12 65	15 73	18 81	22 88	25 96	28 104	31 112	34 120	37 128	41 135	44 143	47 151	50 159	53 167
12	–	–	2 34	5 43	8 52	11 61	14 70	17 79	21 87	24 96	28 104	31 113	35 121	38 130	42 138	46 146	49 155	53 163	56 172	60 180
13	–	0 26	2 37	5 47	9 56	12 66	16 75	20 84	23 94	27 103	31 112	35 121	39 130	43 139	47 148	51 157	55 166	59 175	63 184	67 193
14	–	0 28	2 40	6 50	10 60	13 71	17 81	22 90	26 100	30 110	34 120	38 130	43 139	47 149	51 159	56 168	60 178	65 187	69 197	73 207
15	–	0 30	3 42	7 53	11 64	15 75	19 86	24 96	28 107	33 117	37 128	42 138	47 148	51 159	56 169	61 179	66 189	70 200	75 210	80 220
16	–	0 32	3 45	7 57	12 68	16 80	21 91	26 102	31 113	36 124	41 135	46 146	51 157	56 168	61 179	66 190	71 201	76 212	82 222	87 233
17	–	0 34	4 47	8 60	13 72	18 84	23 96	28 108	33 120	38 132	44 143	49 155	55 166	60 178	66 189	71 201	77 212	82 224	88 234	93 247
18	–	0 36	4 50	9 63	14 76	19 89	24 102	30 114	36 126	41 139	47 151	53 163	59 175	65 187	70 200	76 212	82 224	88 236	94 248	100 260
19	–	1 37	4 53	9 67	15 80	20 94	26 107	32 120	38 133	44 146	50 159	56 172	63 184	69 197	75 210	82 222	88 235	94 248	101 260	107 273
20	–	1 39	5 55	10 70	16 84	22 98	28 112	34 126	40 140	47 153	53 167	60 180	67 193	73 207	80 220	87 233	93 247	100 260	107 273	114 286

(Dashes in the body of the table indicate that no decision is possible at the stated level of significance.)

Table 11. Critical values of T at various levels of probability (Wilcoxon matched-pairs signed-rank test).

	Level of significance for one-tail test			
	.05	.025	.01	.005
	Level of significance for two-tail test			
N	.10	.05	.02	.01
5	0	–	–	–
6	2	0	–	–
7	3	2	0	–
8	5	3	1	0
9	8	5	3	1
10	10	8	5	3
11	13	10	7	5
12	17	13	9	7
13	21	17	12	9
14	25	21	15	12
15	30	25	19	15
16	35	29	23	19
17	41	34	27	23
18	47	40	32	27
19	53	46	37	32
20	60	52	43	37
21	67	58	49	42
22	75	65	55	48
23	83	73	62	54
24	91	81	69	61
25	100	89	76	68
26	110	98	84	75
27	119	107	92	83

	Level of significance for one-tail test			
	.05	.025	.01	.005
	Level of significance for two-tail test			
N	.10	.05	.02	.01
28	130	116	101	91
29	140	126	110	100
30	151	137	120	109
31	163	147	130	118
32	175	159	140	128
33	187	170	151	138
34	200	182	162	148
35	213	195	173	159
36	227	208	185	171
37	241	221	198	182
38	256	235	211	194
39	271	249	224	207
40	286	264	238	220
41	302	279	252	233
42	319	294	266	247
43	336	310	281	261
44	353	327	296	276
45	371	343	312	291
46	389	361	328	307
47	407	378	345	322
48	426	396	362	339
49	446	415	379	355
50	466	434	397	373

Source: *Statistics – An Interdisciplinary Approach* 1974, Open University.

APPENDIX C

Table 12a. Logarithms

											Mean differences								
	0	1	2	3	4	5	6	7	8	9	1	2	3	4	5	6	7	8	9
10	0000	0043	0086	0128	0170	0212	0253	0294	0334	0374	4	8	12	17	21	25	29	33	37
11	0414	0453	0492	0531	0569	0607	0645	0682	0719	0755	4	8	11	15	19	23	26	30	34
12	0792	0828	0864	0899	0934	0969	1004	1038	1072	1106	3	7	10	14	17	21	24	28	31
13	1139	1173	1206	1239	1271	1303	1335	1367	1399	1430	3	6	10	13	16	19	23	26	29
14	1461	1492	1523	1553	1584	1614	1644	1673	1703	1732	3	6	9	12	15	18	21	24	27
15	1761	1790	1818	1847	1875	1903	1931	1959	1987	2014	3	6	8	11	14	17	20	22	25
16	2041	2068	2095	2122	2148	2175	2201	2227	2253	2279	3	5	8	11	13	16	18	21	24
17	2304	2330	2355	2380	2405	2430	2455	2480	2504	2529	2	5	7	10	12	15	17	20	22
18	2553	2577	2601	2625	2648	2672	2695	2718	2742	2765	2	5	7	9	12	14	16	19	21
19	2788	2810	2833	2856	2878	2900	2923	2945	2967	2989	2	4	7	9	11	13	16	18	20
20	3010	3032	3054	3075	3096	3118	3139	3160	3181	3201	2	4	6	8	11	13	15	17	19
21	3222	3243	3263	3284	3304	3324	3345	3365	3385	3404	2	4	6	8	10	12	14	16	18
22	3424	3444	3464	3483	3502	3522	3541	3560	3579	3598	2	4	6	8	10	12	14	15	17
23	3617	3636	3655	3674	3692	3711	3729	3747	3766	3784	2	4	6	7	9	11	13	15	17
24	3802	3820	3838	3856	3874	3892	3909	3927	3945	3962	2	4	5	7	9	11	12	14	16
25	3979	3997	4014	4031	4048	4065	4082	4099	4116	4133	2	3	5	7	9	10	12	14	15
26	4150	4166	4183	4200	4216	4232	4249	4265	4281	4298	2	3	5	7	8	10	11	13	15
27	4314	4330	4346	4362	4378	4393	4409	4425	4440	4456	2	3	5	6	8	9	11	13	14
28	4472	4487	4502	4518	4533	4548	4564	4579	4594	4609	2	3	5	6	8	9	11	12	14
29	4624	4639	4654	4669	4683	4698	4713	4728	4742	4757	1	3	4	6	7	9	10	12	13
30	4771	4786	4800	4814	4829	4843	4857	4871	4886	4900	1	3	4	6	7	9	10	11	13
31	4914	4928	4942	4955	4969	4983	4997	5011	5024	5038	1	3	4	6	7	8	10	11	12
32	5051	5065	5079	5092	5105	5119	5132	5145	5159	5172	1	3	4	5	7	8	9	11	12
33	5185	5198	5211	5224	5237	5250	5263	5276	5289	5302	1	3	4	5	6	8	9	10	12
34	5315	5328	5340	5353	5366	5378	5391	5403	5416	5428	1	3	4	5	6	8	9	10	11
35	5441	5453	5465	5478	5490	5502	5514	5527	5539	5551	1	2	4	5	6	7	9	10	11
36	5563	5575	5587	5599	5611	5623	5635	5647	5658	5670	1	2	4	5	6	7	8	10	11
37	5682	5694	5705	5717	5729	5740	5752	5763	5775	5786	1	2	3	5	6	7	8	9	10
38	5798	5809	5821	5832	5843	5855	5866	5877	5888	5899	1	2	3	5	6	7	8	9	10
39	5911	5922	5933	5944	5955	5966	5977	5988	5999	6010	1	2	3	4	5	7	8	9	10
40	6021	6031	6042	6053	6064	6075	6085	6096	6107	6117	1	2	3	4	5	6	8	9	10
41	6128	6138	6149	6160	6170	6180	6191	6201	6212	6222	1	2	3	4	5	6	7	8	9
42	6232	6243	6253	6263	6274	6284	6294	6304	6314	6325	1	2	3	4	5	6	7	8	9
43	6335	6345	6355	6365	6375	6385	6395	6405	6415	6425	1	2	3	4	5	6	7	8	9
44	6435	6444	6454	6464	6474	6484	6493	6503	6513	6522	1	2	3	4	5	6	7	8	9
45	6532	6542	6551	6561	6571	6580	6590	6599	6609	6618	1	2	3	4	5	6	7	8	9
46	6628	6637	6646	6656	6665	6675	6684	6693	6702	6712	1	2	3	4	5	6	7	7	8
47	6721	6730	6739	6749	6758	6767	6776	6785	6794	6803	1	2	3	4	5	5	6	7	8
48	6812	6821	6830	6839	6848	6857	6866	6875	6884	6893	1	2	3	4	4	5	6	7	8
49	6902	6911	6920	6928	6937	6946	6955	6964	6972	6981	1	2	3	4	4	5	6	7	8
50	6990	6998	7007	7016	7024	7033	7042	7050	7059	7067	1	2	3	3	4	5	6	7	8
51	7076	7084	7093	7101	7110	7118	7126	7135	7143	7152	1	2	3	3	4	5	6	7	8
52	7160	7168	7177	7185	7193	7202	7210	7218	7226	7235	1	2	2	3	4	5	6	7	7
53	7243	7251	7259	7267	7275	7284	7292	7300	7308	7316	1	2	2	3	4	5	6	6	7
54	7324	7332	7340	7348	7356	7364	7372	7380	7388	7396	1	2	2	3	4	5	6	6	7

Source: *Science Data Book*. R.M. Tennant, Oliver & Boyd.

Table 12b. Logarithms

	0	1	2	3	4	5	6	7	8	9	Mean differences 1	2	3	4	5	6	7	8	9
55	7404	7412	7419	7427	7435	7443	7451	7459	7466	7474	1	2	2	3	4	5	5	6	7
56	7482	7490	7497	7505	7513	7520	7528	7536	7543	7551	1	2	2	3	4	5	5	6	7
57	7559	7566	7574	7582	7589	7597	7604	7612	7619	7627	1	2	2	3	4	5	5	6	7
58	7634	7642	7649	7657	7664	7672	7679	7686	7694	7701	1	1	2	3	4	4	5	6	7
59	7709	7716	7723	7731	7738	7745	7752	7760	7767	7774	1	1	2	3	4	4	5	6	7
60	7782	7789	7796	7803	7810	7818	7825	7832	7839	7846	1	1	2	3	4	4	5	6	6
61	7853	7860	7868	7875	7882	7889	7896	7903	7910	7917	1	1	2	3	4	4	5	6	6
62	7924	7931	7938	7945	7952	7959	7966	7973	7980	7987	1	1	2	3	3	4	5	6	6
63	7993	8000	8007	8014	8021	8028	8035	8041	8048	8055	1	1	2	3	3	4	5	5	6
64	8062	8069	8075	8082	8089	8096	8102	8109	8116	8122	1	1	2	3	3	4	5	5	6
65	8129	8136	8142	8149	8156	8162	8169	8176	8182	8189	1	1	2	3	3	4	5	5	6
66	8195	8202	8209	8215	8222	8228	8235	8241	8248	8254	1	1	2	3	3	4	5	5	6
67	8261	8267	8274	8280	8287	8293	8299	8306	8312	8319	1	1	2	3	3	4	5	5	6
68	8325	8331	8338	8344	8351	8357	8363	8370	8376	8382	1	1	2	3	3	4	4	5	6
69	8388	8395	8401	8407	8414	8420	8426	8432	8439	8445	1	1	2	2	3	4	4	5	6
70	8451	8457	8463	8470	8476	8482	8488	8494	8500	8506	1	1	2	2	3	4	4	5	6
71	8513	8519	8525	8531	8537	8543	8549	8555	8561	8567	1	1	2	2	3	4	4	5	5
72	8573	8579	8585	8591	8597	8603	8609	8615	8621	8627	1	1	2	2	3	4	4	5	5
73	8633	8639	8645	8651	8657	8663	8669	8675	8681	8686	1	1	2	2	3	4	4	5	5
74	8692	8698	8704	8710	8716	8722	8727	8733	8739	8745	1	1	2	2	3	4	4	5	5
75	8751	8756	8762	8768	8774	8779	8785	8791	8797	8802	1	1	2	2	3	3	4	5	5
76	8808	8814	8820	8825	8831	8837	8842	8848	8854	8859	1	1	2	2	3	3	4	5	5
77	8865	8871	8876	8882	8887	8893	8899	8904	8910	8915	1	1	2	2	3	3	4	4	5
78	8921	8927	8932	8938	8943	8949	8954	8960	8965	8971	1	1	2	2	3	3	4	4	5
79	8976	8982	8987	8993	8998	9004	9009	9015	9020	9025	1	1	2	2	3	3	4	4	5
80	9031	9036	9042	9047	9053	9058	9063	9069	9074	9079	1	1	2	2	3	3	4	4	5
81	9085	9090	9096	9101	9106	9112	9117	9122	9128	9133	1	1	2	2	3	3	4	4	5
82	9138	9143	9149	9154	9159	9165	9170	9175	9180	9186	1	1	2	2	3	3	4	4	5
83	9191	9196	9201	9206	9212	9217	9222	9227	9232	9238	1	1	2	2	3	3	4	4	5
84	9243	9248	9253	9258	9263	9269	9274	9279	9284	9289	1	1	2	2	3	3	4	4	5
85	9294	9299	9304	9309	9315	9320	9325	9330	9335	9340	1	1	2	2	3	3	4	4	5
86	9345	9350	9355	9360	9365	9370	9375	9380	9385	9390	1	1	2	2	3	3	4	4	5
87	9395	9400	9405	9410	9415	9420	9425	9430	9435	9440	0	1	1	2	2	3	3	4	4
88	9445	9450	9455	9460	9465	9469	9474	9479	9484	9489	0	1	1	2	2	3	3	4	4
89	9494	9499	9504	9509	9513	9518	9523	9528	9533	9538	0	1	1	2	2	3	3	4	4
90	9542	9547	9552	9557	9562	9566	9571	9576	9581	9586	0	1	1	2	2	3	3	4	4
91	9590	9595	9600	9605	9609	9614	9619	9624	9628	9633	0	1	1	2	2	3	3	4	4
92	9638	9643	9647	9653	9657	9661	9666	9671	9675	9680	0	1	1	2	2	3	3	4	4
93	9685	9689	9694	9699	9703	9708	9713	9717	9722	9727	0	1	1	2	2	3	3	4	4
94	9731	9736	9741	9745	9750	9754	9759	9763	9768	9773	0	1	1	2	2	3	3	4	4
95	9777	9782	9786	9791	9795	9800	9805	9809	9814	9818	0	1	1	2	2	3	3	4	4
96	9823	9827	9832	9836	9841	9845	9850	9854	9859	9863	0	1	1	2	2	3	3	4	4
97	9868	9872	9877	9881	9886	9890	9894	9899	9903	9908	0	1	1	2	2	3	3	4	4
98	9912	9917	9921	9926	9930	9934	9939	9943	9948	9952	0	1	1	2	2	3	3	4	4
99	9956	9961	9965	9969	9974	9978	9983	9987	9991	9996	0	1	1	2	2	3	3	3	4

Table 13a. Antilogarithms

											Mean differences								
	0	1	2	3	4	5	6	7	8	9	1	2	3	4	5	6	7	8	9
.00	1000	1002	1005	1007	1009	1012	1014	1016	1019	1021	0	0	1	1	1	1	2	2	2
.01	1023	1026	1028	1030	1033	1035	1038	1040	1042	1045	0	0	1	1	1	1	2	2	2
.02	1047	1050	1052	1054	1057	1059	1062	1064	1067	1069	0	0	1	1	1	1	2	2	2
.03	1072	1074	1076	1079	1081	1084	1086	1089	1091	1094	0	0	1	1	1	1	2	2	2
.04	1096	1099	1102	1104	1107	1109	1112	1114	1117	1119	0	1	1	1	1	2	2	2	2
.05	1122	1125	1127	1130	1132	1135	1138	1140	1143	1146	0	1	1	1	1	2	2	2	2
.06	1148	1151	1153	1156	1159	1161	1164	1167	1169	1172	0	1	1	1	1	2	2	2	2
.07	1175	1178	1180	1183	1186	1189	1191	1194	1197	1199	0	1	1	1	1	2	2	2	2
.08	1202	1205	1208	1211	1213	1216	1219	1222	1225	1227	0	1	1	1	1	2	2	2	3
.09	1230	1233	1236	1239	1242	1245	1247	1250	1253	1256	0	1	1	1	1	2	2	2	3
.10	1259	1262	1265	1268	1271	1274	1276	1279	1282	1285	0	1	1	1	1	2	2	2	3
.11	1288	1291	1294	1297	1300	1303	1306	1309	1312	1315	0	1	1	1	2	2	2	2	3
.12	1318	1321	1324	1327	1330	1334	1337	1340	1343	1346	0	1	1	1	2	2	2	2	3
.13	1349	1352	1355	1358	1361	1365	1368	1371	1374	1377	0	1	1	1	2	2	2	3	3
.14	1380	1384	1387	1390	1393	1396	1400	1403	1406	1409	0	1	1	1	2	2	2	3	3
.15	1413	1416	1419	1422	1426	1429	1432	1435	1439	1442	0	1	1	1	2	2	2	3	3
.16	1445	1449	1452	1455	1459	1462	1466	1469	1472	1476	0	1	1	1	2	2	2	3	3
.17	1479	1483	1486	1489	1493	1496	1500	1503	1507	1510	0	1	1	1	2	2	2	3	3
.18	1514	1517	1521	1524	1528	1531	1535	1538	1542	1545	0	1	1	1	2	2	2	3	3
.19	1549	1552	1556	1560	1563	1567	1570	1574	1578	1581	0	1	1	1	2	2	3	3	3
.20	1585	1589	1592	1596	1600	1603	1607	1611	1614	1618	0	1	1	1	2	2	3	3	3
.21	1622	1626	1629	1633	1637	1641	1644	1648	1652	1656	0	1	1	2	2	2	3	3	3
.22	1660	1663	1667	1671	1675	1679	1683	1687	1690	1694	0	1	1	2	2	2	3	3	3
.23	1698	1702	1706	1710	1714	1718	1722	1726	1730	1734	0	1	1	2	2	2	3	3	4
.24	1738	1742	1746	1750	1754	1758	1762	1766	1770	1774	0	1	1	2	2	2	3	3	4
.25	1778	1782	1786	1791	1795	1799	1803	1807	1811	1816	0	1	1	2	2	2	3	3	4
.26	1820	1824	1828	1832	1837	1841	1845	1849	1854	1858	0	1	1	2	2	3	3	3	4
.27	1862	1866	1871	1875	1879	1884	1888	1892	1897	1901	0	1	1	2	2	3	3	3	4
.28	1905	1910	1914	1919	1923	1928	1932	1936	1941	1945	0	1	1	2	2	3	3	4	4
.29	1950	1954	1959	1963	1968	1972	1977	1982	1986	1991	0	1	1	2	2	3	3	4	4
.30	1995	2000	2004	2009	2014	2018	2023	2028	2032	2037	0	1	1	2	2	3	3	4	4
.31	2042	2046	2051	2056	2061	2065	2070	2075	2080	2084	0	1	1	2	2	3	3	4	4
.32	2089	2094	2099	2104	2109	2113	2118	2123	2128	2133	0	1	1	2	2	3	3	4	4
.33	2138	2143	2148	2153	2158	2163	2168	2173	2178	2183	0	1	1	2	2	3	3	4	4
.34	2188	2193	2198	2203	2208	2213	2218	2223	2228	2234	1	1	2	2	3	3	4	4	5
.35	2239	2244	2249	2254	2259	2265	2270	2275	2280	2286	1	1	2	2	3	3	4	4	5
.36	2291	2296	2301	2307	2312	2317	2323	2328	2333	2339	1	1	2	2	3	3	4	4	5
.37	2344	2350	2355	2360	2366	2371	2377	2382	2388	2393	1	1	2	2	3	3	4	4	5
.38	2399	2404	2410	2415	2421	2427	2432	2438	2443	2449	1	1	2	2	3	3	4	4	5
.39	2455	2460	2466	2472	2477	2483	2489	2495	2500	2506	1	1	2	2	3	3	4	5	5
.40	2512	2518	2523	2529	2535	2541	2547	2553	2559	2564	1	1	2	2	3	4	4	5	5
.41	2570	2576	2582	2588	2594	2600	2606	2612	2618	2624	1	1	2	2	3	4	4	5	5
.42	2630	2636	2642	2649	2655	2661	2667	2673	2679	2685	1	1	2	2	3	4	4	5	6
.43	2692	2698	2704	2710	2716	2723	2729	2735	2742	2748	1	1	2	3	3	4	4	5	6
.44	2754	2761	2767	2773	2780	2786	2793	2799	2805	2812	1	1	2	3	3	4	4	5	6
.45	2818	2825	2831	2838	2844	2851	2858	2864	2871	2877	1	1	2	3	3	4	5	5	6
.46	2884	2891	2897	2904	2911	2917	2924	2931	2938	2944	1	1	2	3	3	4	5	5	6
.47	2951	2958	2965	2972	2979	2985	2992	2999	3006	3013	1	1	2	3	3	4	5	5	6
.48	3020	3027	3034	3041	3048	3055	3062	3069	3076	3083	1	1	2	3	4	4	5	6	6
.49	3090	3097	3105	3112	3119	3126	3133	3141	3148	3155	1	1	2	3	4	4	5	6	6

Source: *Science Data Book*, R.M. Tennant, Oliver & Boyd.

Table 13b. Antilogarithms

	0	1	2	3	4	5	6	7	8	9	Mean differences								
											1	2	3	4	5	6	7	8	9
.50	3162	3170	3177	3184	3192	3199	3206	3214	3221	3228	1	1	2	3	4	4	5	6	7
.51	3236	3243	3251	3258	3266	3273	3281	3289	3296	3304	1	2	2	3	4	5	5	6	7
.52	3311	3319	3327	3334	3342	3350	3357	3365	3373	3381	1	2	2	3	4	5	5	6	7
.53	3388	3396	3404	3412	3420	3428	3436	3443	3451	3459	1	2	2	3	4	5	6	6	7
.54	3467	3475	3483	3491	3499	3508	3516	3524	3432	3540	1	2	2	3	4	5	6	6	7
.55	3548	3556	3565	3573	3581	3589	3597	3606	3614	3622	1	2	2	3	4	5	6	7	7
.56	3631	3639	3648	3656	3664	3673	3681	3690	3698	3707	1	2	3	3	4	5	6	7	8
.57	3715	3742	3733	3741	3750	3758	3767	3776	3784	3793	1	2	3	3	4	5	6	7	8
.58	3802	3811	3819	3828	3837	3846	3855	3864	3873	3882	1	2	3	4	4	5	6	7	8
.59	3890	3899	3908	3917	3926	3936	3945	3954	3963	3972	1	2	3	4	5	5	6	7	8
.60	3981	3990	3999	4009	4018	4027	4036	4046	4055	4064	1	2	3	4	5	6	6	7	8
.61	4074	4083	4093	4102	4111	4121	4130	4140	4150	4159	1	2	3	4	5	6	7	8	9
.62	4169	4178	4188	4198	4207	4217	4227	4236	4246	4256	1	2	3	4	5	6	7	8	9
.63	4266	4276	4285	4295	4305	4315	4325	4335	4345	4355	1	2	3	4	5	6	7	8	9
.64	4365	4375	4385	4395	4406	4416	4426	4436	4446	4457	1	2	3	4	5	6	7	8	9
.65	4467	4477	4487	4498	4508	4519	4529	4539	4550	4560	1	2	3	4	5	6	7	8	9
.66	4571	4581	4592	4603	4613	4624	4634	4645	4656	4667	1	2	3	4	5	6	7	9	10
.67	4677	4688	4699	4710	4721	4732	4742	4753	4764	4775	1	2	3	4	5	7	8	9	10
.68	4786	4797	4808	4819	4831	4842	4853	4864	4875	4887	1	2	3	4	6	7	8	9	10
.69	4898	4909	4920	4932	4943	4955	4966	4977	4989	5000	1	2	3	5	6	7	8	9	10
.70	5012	5023	5035	5047	5058	5070	5082	5093	5105	5117	1	2	4	5	6	7	8	9	11
.71	5129	5140	5152	5164	5176	5188	5200	5212	5224	5236	1	2	4	5	6	7	8	10	11
.72	5248	5260	5272	5284	5297	5309	5321	5333	5346	5358	1	2	4	5	6	7	9	10	11
.73	5370	5383	5395	5408	5420	5433	5445	5458	5470	5483	1	3	4	5	6	8	9	10	11
.74	5495	5508	5521	5534	5546	5559	5572	5585	5598	5610	1	3	4	5	6	8	9	10	12
.75	5623	5636	5649	5662	5675	5689	5702	5715	5728	5741	1	3	4	5	7	8	9	10	12
.76	5754	5768	5781	5794	5808	5821	5834	5848	5861	5875	1	3	4	5	7	8	9	11	12
.77	5888	5902	5916	5929	5943	5957	5970	5984	5998	6012	1	3	4	5	7	8	10	11	12
.78	6026	6039	6053	6067	6081	6095	6109	6124	6138	6152	1	3	4	6	7	8	10	11	13
.79	6166	6180	6194	6209	6223	6237	6252	6266	6281	6295	1	3	4	6	7	9	10	11	12
.80	6310	6324	6339	6353	6368	6383	6397	6412	6427	6442	1	3	4	6	7	9	10	12	13
.81	6457	6471	6485	6501	6516	6531	6546	6561	6577	6592	2	3	5	6	8	9	11.	12	14
.82	6607	6622	6637	6653	6668	6683	6699	6714	6730	6745	2	3	5	6	8	9	11	12	14
.83	6761	6776	6792	6808	6823	6839	6855	6871	6887	6902	2	3	5	6	8	9	11	13	14
.84	6918	6934	6950	6966	6982	6998	7015	7031	7047	7063	2	3	5	6	8	10	11	13	15
.85	7079	7096	7112	7129	7145	7161	7178	7194	7211	7228	2	3	5	7	8	10	12	13	15
.86	7244	7261	7278	7295	7311	7328	7345	7362	7379	7396	2	3	5	7	8	10	12	13	15
.87	7413	7430	7447	7464	7482	7499	7516	7534	7551	7568	2	3	5	7	9	10	12	14	16
.88	7586	7603	7621	7638	7656	7674	7691	7709	7727	7745	2	4	5	7	9	11	12	14	16
.89	7762	7780	7798	7816	7834	7852	7870	7889	7907	7925	2	4	5	7	9	11	13	14	16
.90	7943	7962	7980	7998	8017	8035	8054	8072	8091	8110	2	4	6	7	9	11	13	15	17
.91	8128	8147	8166	8185	8204	8222	8241	8260	8279	8299	2	4	6	8	9	11	13	15	17
.92	8318	8337	8356	8375	8395	8414	8433	8453	8472	8492	2	4	6	8	10	12	14	15	17
.93	8511	8531	8551	8570	8590	8610	8630	8650	8670	8690	2	4	6	8	10	12	14	16	18
.94	8710	8730	8750	8770	8790	8810	8831	8851	8872	8892	2	4	6	8	10	12	14	16	18
.95	8913	8933	8954	8974	8995	9016	9036	9057	9078	9099	2	4	6	8	10	12	15	17	19
.96	9120	9141	9162	9183	9204	9226	9247	9268	9290	9311	2	4	6	8	11	13	15	17	19
.97	9333	9354	9376	9397	9419	9441	9462	9484	9506	9528	2	4	7	9	11	13	15	17	20
.98	9550	9572	9594	9616	9638	9661	9683	9705	9727	9750	2	4	7	9	11	13	16	18	20
.99	9772	9795	9817	9840	9863	9886	9908	9931	9954	9977	2	5	7	9	11	14	16	18	20

Table 14. Random numbers

39103	06055	58776	04929	08103
23417	74529	03659	58798	58738
95389	14986	66450	73208	42524
98121	94608	32848	58158	46146
93315	37771	69092	40737	66690
96403	77677	24386	11516	06202
50313	78342	23784	60544	43904
62993	10989	53867	00633	45589
36555	56693	89923	64374	92355
80930	03081	19919	19144	59507
70367	67301	40214	29795	69648
85097	45518	85431	65746	76319
52094	60679	35097	81751	83148
86041	28973	12755	18795	16729
83855	30113	49807	93548	35335
68554	21020	45035	45059	62385
36782	08561	90467	91239	17232
58355	12805	19059	61327	16664
40176	46134	12909	12133	90952
27993	78506	53582	62681	22164
98804	06710	37725	26598	03928
81501	88793	14939	29896	99070
03374	06142	13814	13472	37723
94768	64844	60132	16513	86787
30871	87317	71329	64774	01830
72151	48526	36545	12147	54796
73936	22532	64058	79448	77405
23117	37774	50818	53241	83980
18307	54346	64984	09685	10532
92314	45902	05161	43115	27073
60413	20189	22379	12324	01422
41538	51273	11021	92544	84319
84262	43243	77339	42128	54466
08205	68712	90447	83100	26739
69337	34776	90677	41531	96803
23819	67221	23112	33406	59305
92368	15340	45660	77226	80195
29755	39159	18585	64049	64822
16661	89016	16666	03281	92295
59932	04627	58117	71636	04412
73503	49573	58626	18886	92325
24273	66534	23507	18090	08533
24215	00149	68236	33923	38996
90169	93118	24413	19001	62796
76477	76816	59948	53358	99126
65940	43490	30914	71580	84105
90619	42457	68554	20207	99822
66491	69055	17716	36095	12091
77734	04950	59002	62530	50490
94362	49185	79920	09405	94355

COLLINS
REFERENCE

COLLINS REFERENCE DICTIONARIES form a new series of paperback subject dictionaries covering a wide variety of academic and general topics, ranging from Biology to Business, and from Maths to Music. Collins Reference Dictionaries fully maintain the reputation of Collins English and bilingual dictionaries for clarity, authority, comprehensiveness and ease-of-use.

For the series, Collins have commissioned expert authors who combine a profound understanding of their subject with great skill and experience in explaining and clarifying difficult terms and concepts. A consistent style and approach is adopted throughout the series with encyclopedic-style explanations following each precise initial definition. Where appropriate, extensive use is made of worked examples, line illustrations, graphs and tables to further aid the reader.

The academic titles are primarily designed to meet the needs of undergraduates, although the dictionaries will also be of considerable value to advanced school students and to people working in related disciplines. The general titles will provide the layman with comprehensive, lively and up-to-date reference works for a wide range of subjects.

Details of some of this exciting new range of dictionaries can be found on the following pages.

COLLINS DICTIONARY OF MATHEMATICS

Ephraim Borowski and Jonathan Borwein

Containing over 4,000 entries and 400 diagrams, this is the only reference work at this level to present this coverage at an accessible price. Covering an enormous range of technical terms from both pure and applied maths, the Dictionary goes beyond basic definitions to provide helpful explanations and examples, geared to the level of student who is liable to be looking up the particular term.

At undergraduate level, the reader will find extensive information on such fields as real and complex analysis, abstract algebra, number theory, metamathematics and the foundations of mathematics, topology, vector calculus, continuum mechanics, differential equations, measure theory, and graph theory. At the advanced school level the subjects covered include set theory, matrices, trigonometry and geometry, calculus, mechanics, statistics, and logic. The biographies of major mathematicians and the discursive explanations of paradoxes also make the Dictionary interesting and informative for the browser, and help to round off what will be an ideal course companion at many levels.

Ephraim Borowski, B.Phil., is a lecturer in philosophy at Glasgow University with a particular interest in logic and the philosophy of mathematics. He is also mathematics consultant to the *Collins English Dictionary*. Dr Jonathan Borwein is a professor of mathematics at Dalhousie University, Nova Scotia.

ISBN 0 00 434347 6

COLLINS DICTIONARY OF ECONOMICS

Christopher Pass, Bryan Lowes and Leslie Davies

The *Collins Dictionary of Economics* is a completely new and definitive guide to the intricate structures and forces of the economic world. The Dictionary is addressed primarily to the requirements of undergraduates and advanced school students, but will also prove an invaluable reference book for students of the subject at a wide range of levels or for anyone studying economics as part of a broader-based course such as business studies or social science.

The Dictionary has over 1,700 entries – from *ability-to-pay principle of taxation* to *zero-sum game*. There are also over 190 diagrams, plus numerous tables. Essential mathematical and statistical terms are included, as are brief biographies of major economists.

In addition to providing clear definitions of economic terms, the Dictionary summarizes the important theoretical principles behind the science of economics, from the "invisible hand" of Smith and the class analysis of Marx to the "rigid wages" of Keynes and the free-market monetarism of Friedman.

As well as the basic definition and explanation of a particular term, the reader who requires further exposition is guided, where appropriate, through cross references to related terms and refinements of the original concept.

Dr Pass and Dr Lowes are lecturers in managerial economics at the University of Bradford's Management Centre. Dr Davies is a research assistant at the same institution.

ISBN 0 00 434353 0

COLLINS DICTIONARY OF BIOLOGY

W. G. Hale and J. P. Margham

The *Collins Dictionary of Biology* is a completely new, authoritative and comprehensive guide to the complex discipline of biology. The book is directed primarily to the needs of undergraduates and advanced school students, but will also prove an invaluable source book for anyone with an interest in this wide-ranging and fascinating subject.

The Dictionary has over 5,600 entries ranging from *abdomen* to *zymogen*. There are also 285 diagrams illustrating such features as the generalized structures of animals, genetic organization, parts of plants and animals, etc., plus numerous tables.

All the major fields within biology are covered, including anatomy, biochemistry, ecology, genetics, physiology, evolutionary theory and taxonomy. The entries have been structured to provide more than an isolated definition of the term in question. An entry such as *acquired characters* is thus set in the context of the development of evolutionary theory, while the series of genetic entries guides the reader through such concepts as the basis of *genetic code* and the industrial implications of *genetic engineering*. The Dictionary also includes brief biographies of major biologists.

Professor Hale is Dean of the Faculty of Science and Head of the Department of Biology at Liverpool Polytechnic. Dr Margham is Principal Lecturer and Course Leader for the B.Sc. Honours Applied Biology Degree at Liverpool Polytechnic.

ISBN 0 00 434351 4

COLLINS DICTIONARY OF ELECTRONICS

Ian R. Sinclair

The *Collins Dictionary of Electronics* is a completely new and up-to-date guide to the science and technology of electronics. Containing over 2,000 entries, from *aberration* to *zero error*, the Dictionary also includes over 100 diagrams, together with lists of symbols used in electronics.

The Dictionary is intended for anyone who needs a source book providing clear, helpful definitions of electronic terms, including advanced school students and those embarking on higher-education courses, as well as technicians and hobbyists. The Dictionary will also prove useful to anyone whose work or study involves the use of electronic devices – which now have become vital tools in areas as diverse as music, archaeology and medicine.

The Dictionary guides the reader through the various fields within electronics such as microprocessor technology, digital electronics, telecommunications, hi-fi, radio, and television. The emphasis throughout is on the practical application of concepts and devices, although the theoretical background is also well covered, and the reader will find entries on such topics as the *superposition theorem*, the *Biot-Savart law*, and the *permeability* and *permittivity of free space*. Where appropriate, the mathematical aspect of a topic is introduced, although this is generally avoided.

Ian R. Sinclair is a professional technical author. He has written many books on computing for Collins, including the *Collins Dictionary of Computing*.

ISBN 0 00 434345 X

COLLINS DICTIONARY OF COMPUTING

Ian R. Sinclair

Aimed primarily at those using microcomputers as a tool – whether it be at university, school or college, or in the home or office – rather than the professional computer expert, the Dictionary will also prove invaluable to anyone whose work is related in any way to computers, be they micro, mini or mainframe.

Far too many computer manuals assume that even the beginner knows the meaning of a vast array of specialized jargon, leaving the user in a state of near despair. Although not a handbook to particular machines, *Collins Dictionary of Computing* contains definitions and explanations of over 2,000 of the terms that the average user is most likely to come across, from *access* to *zero compression*. Written with a clarity and precision that will be welcomed by all computer users, the entries are augmented by nearly 100 diagrams and explanatory captions.

Comprehensive without indulging in unnecessary padding, and of course completely up-to-date, the Dictionary includes such basic terms as *BASIC*, *bit* and *binary*, as well as those strange terms that computer buffs have made peculiarly their own, such as *blow*, *bomb* and *bubble*. Fields covered range through hardware, software, programming, computer logic, data and word processing, languages, systems, and graphics, to those areas of information technology in which computers play such a vital role.

Ian Sinclair, a full-time technical author, has written numerous best-selling computer books for Collins, and is also author of the *Collins Dictionary of Electronics*.

ISBN 0 00 434349 2

COLLINS DICTIONARY OF MUSIC

Sir Jack Westrup and F. L. Harrison
Revised by Conrad Wilson

Collins Dictionary of Music is a concise third edition of the critically acclaimed *Collins Encyclopedia of Music*, and brings that work right up to date with the inclusion of entries on such figures as Jessye Norman, Michael Berkeley, Arvo Pärt, Nigel Osborne, Janet Glover, Philip Glass, Steve Reich and Nigel Kennedy. In addition, existing entries have been updated with, for example, the inclusion of composers' new works, and details of performers' recent careers.

Covering the whole field of Western music, over 6,500 articles describe composers, instruments, compositions, technical terms, performers, musical forms, periods, styles, movements, critics, musicologists, librettists, instrument makers, orchestras, and opera companies.

The essay-length articles on the major composers include a biographical outline and an assessment of their achievements, together with a list of their principal compositions. Extensive articles have also been devoted to such topics as the history of opera, ballet, film music, and music criticism. The numerous articles on musical theory and notation, carefully cross-referenced, are clarified by the use of music examples, and there is also an appendix of signs and symbols.

The late Sir Jack Westrup was Oxford Professor of Music and one of the foremost musical scholars of his day. Frank Harrison has had a distinguished teaching career, his posts having included professorships at Amsterdam, Yale and Princeton. Conrad Wilson, who also revised the second edition, is music critic of *The Scotsman*.

ISBN 0 00 434356 5

COLLINS DICTIONARY OF ART AND ARTISTS

General Editor: Sir David Piper

Collins Dictionary of Art and Artists is the most up-to-date and comprehensive dictionary of art available in paperback. It covers all schools and periods of Western art – from the Ancient Greeks to the present day – in almost 3,000 entries. As well as the giants of painting and sculpture, the Dictionary also includes many fascinating minor figures, and in addition there are entries on groups, movements, writers on art, materials, and techniques.

Contributed by a panel of more than 70 scholars under the General Editorship of Sir David Piper, the entries are succinct, accurate and lively, forming an authoritative guide to all the names and terms the general reader might wish to look up. Entries range in length from a few lines up to 700 words, and although they are concise they do much more than simply list facts. Artists are firmly characterized so as to give the reader a clear idea not only of their work but also of their significance in the history of art. Thousands of works of art are cited, with dates and locations, and the text is enlivened with contemporary anecdotes and quotations.

Sir David Piper is Director of the Ashmolean Museum in Oxford, and one of the most distinguished art historians and critics of his generation. His numerous publications range from the scholarly to the popular, and he has reached a wide audience as one of the presenters of the BBC television series *100 Great Paintings*.

ISBN 0 00 434358 1